Mathematical Series 1

Lecture notes on
$\mathbb{Z}$ -Transform
First Edition

Refaat El Attar
Professor of Mathematics
College of Engineering
Alexandria University
Egypt
Rea5@hotmail.com

Refaat El Attar

Lulu Press Incorporated
3131 RDU Center Suite 210
Morrisville NC 27560 USA
Http://www.lulu.com

ISBN : 1-4116-1979-X
EAN : 978-1-4116-1979-1

Printed in the United States of America

PREFACE

$\mathbb{Z}$-Transform is one of several transforms that are essential mathematical tools used in engineering and applied sciences. This edition of this book was written to provide an introduction to the subject of $\mathbb{Z}$-Transform. The material presented in this book can be covered in four to five 2-hour classroom lectures. Basic knowledge of calculus is needed. The book is intended to help readers and students in engineering and applied sciences understand the basic properties of $\mathbb{Z}$-Transform and some of the methods and techniques based on this transform to solve some engineering and science problems.

I have collected many examples and problems on the subject that might help the reader getting on-hand experience with the techniques presented in this note.

Finally, I express my appreciation to Dr. Soraya Masoud and Mina A. Makar for their cooperation during the preparation of this note.

Refaat A. El Attar
Alexandria December 1, 2005

Refaat El Attar

IV

CONTENTS

Page

Chapter One

The ℤ -Transform

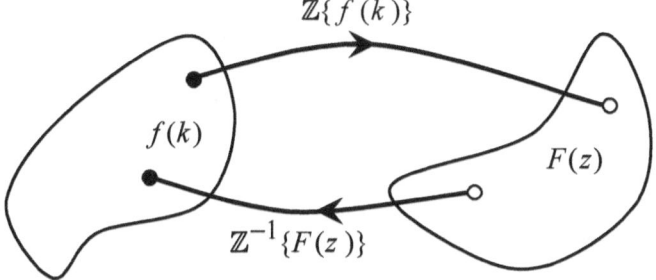

Refaat El Attar

Chapter 1

The $\mathbb{Z}$ -Transform

1.1. Introduction

The operation of a continuous-time system is described or modeled by a set of differential equations. On the other hand, a discrete-time system (or a sampled-data system) is described by a set of difference equations. The transform method employed in the analysis of continuous-time systems is the Laplace transformation. In a similar manner, the transform used in the analysis of discrete-time systems is the $\mathbb{Z}$ -Transform. In other words, $\mathbb{Z}$ -Transform is the discrete counterpart of the Laplace Transform. Also, it may be thought as a generalization of the Fourier Transform for sampled-data signal as we shall see later on.

We all know that the use of transforms in mathematical treatment to problems arising in certain applications is very important. The use of transforms has many advantages and because of these advantages, they are considered very powerful mathematical tools. These advantages are:

1. Turning the differential or difference equations to algebraic ones, which are easier to solve.

2. The involved operation of convolution in the time domain is reduced to a multiplication operation in the transform domain.

3. The initial conditions are directly incorporated in the solution process and do not have to be separately incorporated.

4. The representation of a system in terms of the locations of poles and the zeros of the system transfer function in the complex plane.

5. The transient and steady state characteristics of a discrete system can be obtained by analyzing the poles and zeros of the system.

The idea of transforms is multiplying the function in the time domain by another function called the Kernel, which is a function of both the time and the transform domain variable. Then we perform a definite integral with respect to time and thus we have a function of the transform domain variable. The resulting function is equivalent to the time domain function but in the transform domain.

1.2. Discrete-Time System

A digital computer used as a controller in a feedback control system requires an input signal in digital form. However, the plant being controlled will generally be providing signals that vary continuously with time, i.e. they are analog signals. A digital thermometer is a device that gives a digital output for an analog input, the temperature being an analog quantity that varies continuously with time. With both these examples, the input signals have to be converted from analog to digital form. Such a conversion requires the signal to be sampled at intervals, each sample then being converted into digital form.

1.2.1. The Sampling Process

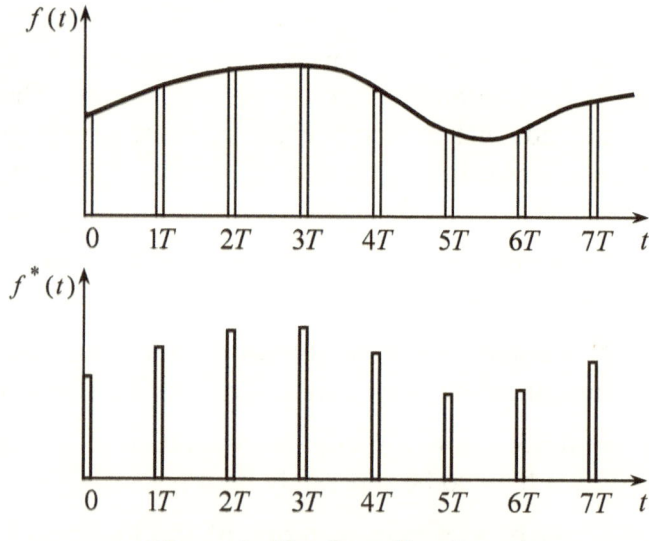

Figure 1. The Sampling Process

Consider a continuous input function $f(t)$. Samples of the input waveform are taken every T seconds. The sampling element senses the value of the signal over a short interval of time Δt in each sampling period and ignores it for the rest of the period. Figure 1 illustrates the described sampling process for the continuous-time signal $f(t)$. The output $f^*(t)$, generally referred as $f(k)$, is a series of pulses at time $0, T, 2T, ..., kT, ...$, being a measure of the size of the continuous-time signal $f(t)$ at that time. We can represent the pulses as the sequence

$$f(0), f(T), f(2T), ..., f(kT), ...$$

where k is an integer number, or we can simply use the notation

$$\{f(k)\}_{k=0}^{\infty} = \{f(0), f(1), f(2), ..., f(k), ...\}$$

1.2.2. Examples:

1. Consider an analog signal input that is just a unit impulse at $t = 0$, the sampled output is represented by Figure 2a, the series of pulses is $1, 0, 0, ..., 0, ...$, which can be denoted by $\delta(k)$.

2. For an input of a unit impulse at $t = 2T$, the output series is shown in Figure 2b and is $0, 0, 1, 0, 0, ..., 0, ...$, which can be denoted by $\delta(k-2)$.

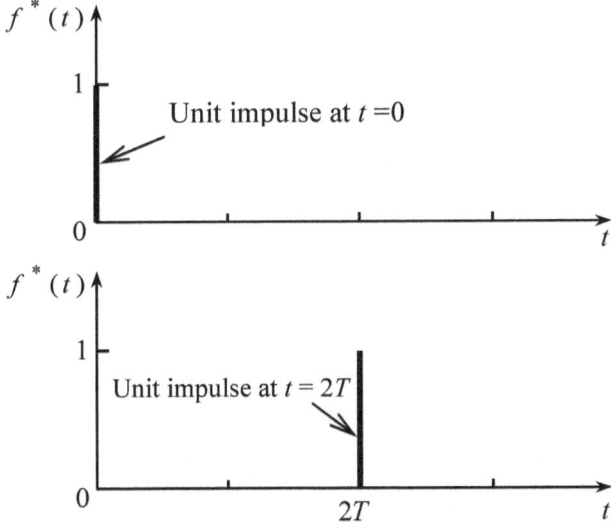

Figure 2. Unit impulses at (a) $t = 0$, (b) $t = 2T$

3. For an input of a unit step at $t = 0$, the series of pulses produced is as shown in Figure 3, and is $1, 1, 1, ..., 1, ...$, which can be denoted by $u(k)$.

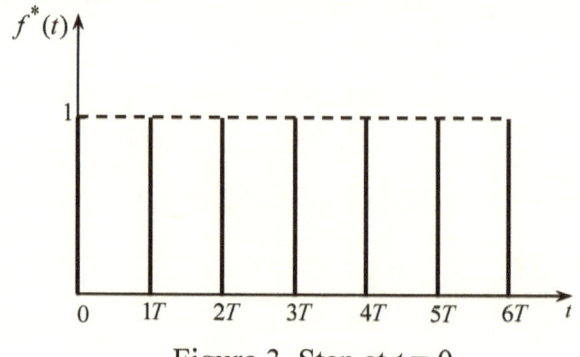

Figure 3. Step at $t = 0$

4. For an input of a ramp $f(t) = t$, the series of pulses produced is as shown in Figure 4, and is $0, 1, 2, 3, ..., k, ...$

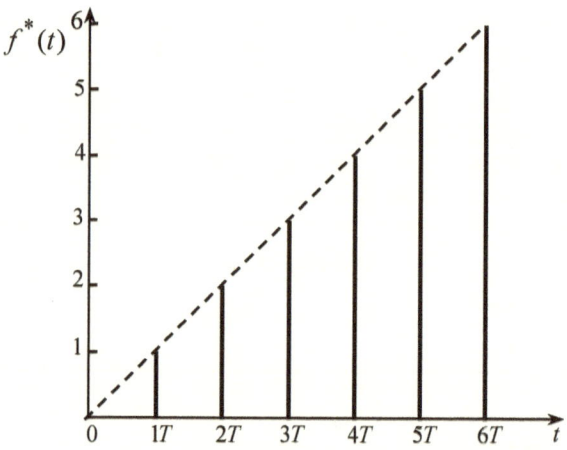

Figure 4. Ramp starting at $t = 0$

Whatever the form of the continuous function, the digital output is a sequence of impulses. Each impulse in the sequence is a unit impulse multiplied by the value of the continuous function $f(t)$ at that time.

We can write $f^*(t)$ that describes the sequence of pulses for a function $f(t)$ with a sampling period T as

$$f^*(t) = f(0) + f(T)\delta(t - T) + f(2T)\delta(t - 2T)$$

$$+ ... + f(kT)\delta(t - kT) + ... \qquad (1)$$

1.2.3. Example on varying the sampling period

Varying the sampling interval will affect the resulting sequence. Consider the continuous signal $f(t) = 2t$, what will be sampled output from this analog input signal, when the sampling period is: (a) 1 second; (b) 0.5 second?

The sequence corresponding to the analog input signal is $\{2kT\}_{k=0}^{\infty} = \{0, 2T, 4T, ..., 2kT, ...\}$. Then we have:

(a) For $T = 1$ second, the sequence is

$$\{2k\}_{k=0}^{\infty} = \{0, 2, 4, ..., 2k, ..\}.$$

(b) For $T = 0.5$ second, the sequence is

$$\{k\}_{k=0}^{\infty} = \{0, 1, 2, ..., k, ...\}.$$

For the sake of simplifying the analysis, we will restrict the sampling period to the value of one. In this case we will denote a sequence in the time domain by $\{f(k)\}$, where k is an integer number.

Sequences can be finite or infinite. Infinite sequences take the form

$$\{f(k)\}_{k=0}^{\infty} = \{f(0), f(1), f(2), ..., f(k), ...\}.$$

The sequence $\{f(k)\}_{k=0}^{\infty} = \{1, 4, 7, 10, 13, ...\}$, is an arithmetic sequence, whose k^{th} term is written as $f(k) = 1 + 3k$. Clearly, the first number of this sequence is $f(0) = 1$. The same information for this sequence can be given as

$$f(k+1) = f(k) + 3, \quad f(0) = 1.$$

This type of representation is exactly what a difference equation is all about. It is a recursion formula that relates the value at time $k+1$ to that at time k. This relation is a first order difference equation with initial condition $f(0) = 1$.

1.3. The $\mathbb{Z}$-Transform

The $\mathbb{Z}$-transform is a mathematical transform with many properties similar to those of the Laplace and Fourier transforms. The main difference is that it operates on a sequence $\{f(k)\}$ of the discrete integer valued argument k rather than on a piecewise continuous function as in the case of Laplace or Fourier transforms.

1.3.1. Definition of the $\mathbb{Z}$-Transform

The $\mathbb{Z}$-transform is an operator that maps a sequence $f(k)$, which is an element of the original space (e.g. the time domain), into a function $F(z)$, (where z is a complex variable). The function $F(z)$ is an element of the transform space (e.g. the complex or frequency domain).

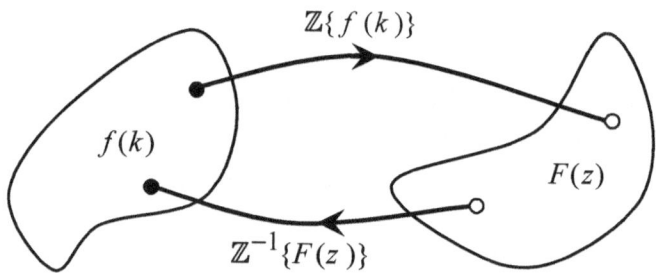

The transformation is written as $\mathbb{Z}\{f(k)\} = F(z)$ whereas the inverse transformation is $\mathbb{Z}^{-1}\{F(z)\} = f(k)$. The $\mathbb{Z}$-transform of a sequence $\{f(k)\}$ is defined as a function $F(z)$ of the complex variable z by the infinite series

$$\mathbb{Z}\{f(k)\} = F(z) = \sum_{k=0}^{\infty} f(k)z^{-k}$$

This is often referred to as the **unilateral** $\mathbb{Z}$-transform. At times the $\mathbb{Z}$-transform is defined as

$$\mathbb{Z}\{f(k)\} = F(z) = \sum_{k=-\infty}^{\infty} f(k)z^{-k}$$

This is the *bilateral* ℤ -transform.

The Fourier transform can be obtained from the ℤ -transform by making the substitution $z = e^{j\omega}$, note that this corresponds to restricting $|z| = 1$, then

$$F(e^{j\omega}) = \sum_{k=-\infty}^{\infty} f(k)(e^{j\omega})^{-k} = \sum_{k=-\infty}^{\infty} f(k)e^{-j\omega k}.$$

This is the Fourier transform of the sequence $\{f(k)\}$. The Fourier transform corresponds to the ℤ -transform evaluated on the unit circle.

In this book, any reference to ℤ -transform is meant the unilateral case unless precisely specified. Note that a signal in the time domain from 0 to ∞ is called a *causal signal* while a signal in the time domain from - ∞ to 0 is called a *non-causal* (anti-causal) *signal*.

The region of convergence, often known as the **ROC**, is important to understand because it defines the region where the ℤ -transform exists. The **ROC** of a sequence $\{f(k)\}$ is defined as the range of z for which the sequence generated by the application of the ℤ -transform converges. Since the ℤ -transform is a *power series*, it converges when $f(k)z^{-k}$ is absolutely summable, *i.e.*

$$\sum_{k=0}^{\infty} \left| f(k) z^{-k} \right| < \infty,$$

must be satisfied for convergence. The region of convergence will be discussed in details later on.

Example: Given $\{x(k)\} = \{2, 4, 6, 4, 2\}$ for values of $0 \le k \le 4$, find $X(z)$.

Solution: From the basic definition

$$X(z) = \sum_{k=0}^{4} x(k) z^{-k} = 2z^0 + 4z^{-1} + 6z^{-2} + 4z^{-3} + 2z^{-4} \ \square$$

Example: Given $X(z) = 1 - 2z^{-1} + 3z^{-3} - z^{-5}$, find $\{x(k)\}$

Solution: From the basic definition

$$\{x(k)\} = \{1, -2, 0, 3, 0, -1\}$$

$$\text{or} \quad x(k) = \delta(k) - 2\,\delta(k\text{-}1) + 3\,\delta(k\text{-}3) - \delta(k\text{-}5).$$ □

1.3.2. Derivation of the Transform

The Laplace transform of an impulse at $t = 0$ is 1. The Laplace transform of an impulse at time T is e^{-Ts} and at time $2T$ is e^{-2Ts} and so on. In general the Laplace transform of an impulse at $t = kT$ is e^{-kTs}, thus the Laplace transform of $f^*(t)$ of equation (1) is

$$F(s) = f(0) \cdot 1 + f(T)e^{-Ts} + f(2T)e^{-2Ts}$$
$$+ \ldots + f(kT)e^{-kTs} + \ldots \tag{2}$$

or $\qquad F(s) = \sum_{k=0}^{\infty} f(kT)e^{-kTs} \qquad (3)$

Let $z = e^{Ts}$ or $s = \dfrac{1}{T}\ln z$, then $F(s)$ with s having this value is called the $\mathbb{Z}$-transform of $f^*(t)$ and is written as $\mathbb{Z}\{f^*(t)\} = F(z)$. Since $f^*(t)$ is a sequence of pulses that can be represented by $\{f(k)\}$, then this expression is often written as $\mathbb{Z}\{f(k)\} = F(z)$. Thus equation (3) can be written as

$$F(z) = f(0)z^0 + f(T)z^{-1} + f(2T)z^{-2} + \ldots + f(kT)z^{-k} + \ldots \tag{4}$$

$$= \sum_{k=0}^{\infty} f(kT)z^{-k}$$

In general, if any continuous function has a Laplace transform then the corresponding sampled function has a $\mathbb{Z}$-transform.

1.3.3. Relationship between the *s*-plane and the *z*-plane

As there is a relationship between the Laplace transform and the $\mathbb{Z}$-transform, there must be a relationship between the *s*-plane and the *z*-plane.

To find such relationship, we use the concept of mapping, which is affecting a group of points in one plane with the relationship between the variables of the two planes to see how these points are translated (mapped) in the other plane.

Before proceeding any further, we notice here that the mapping concept and the following relationships are very important in studying different continuous time systems (using Laplace transform) and discrete time systems (using $\mathbb{Z}$-transform).

The mapping concept is illustrated in Figure 5. Some important points in the *s*-plane and their counterparts in the *z*-plane are shown in the Figure.

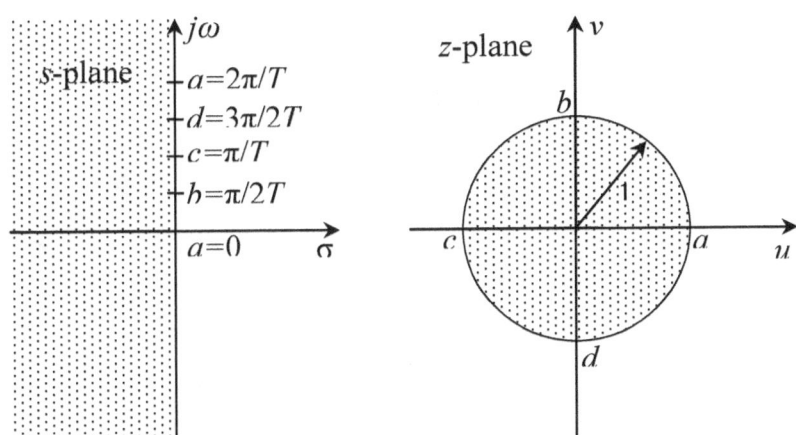

Figure 5. Mapping the *jω*-axis from *s*-plane to *z*-plane

$$z = e^{sT} \ or \ z = e^{(\sigma + j\omega)T} = e^{\sigma T} e^{j\omega T}$$

Mapping the $j\omega$ - axis with $\sigma = 0$ and $s = j\omega$ where

$$z = e^{j\omega T} = \cos \omega T + j \sin \omega T$$

$$= \sqrt{\cos^2 \omega T + \sin^2 \omega T} \ \underline{|\tan^{-1}(\tan \omega T)} = 1 \ \underline{|\omega T} = 1 \ \underline{|\phi}$$

Therefore, points on the $j\omega$–axis in the s-plane are mapped to points on the unit circle in the Z-plane.

For the **Left Hand Plane (LHP)**: $\sigma < 0$, $z = e^{\sigma T} e^{j\omega T}$; therefore points in LHP in the s-plane are mapped to points inside the unit circle in the Z-plane.

For the **Right Hand Plane (RHP)**: $\sigma > 0$, $z = e^{\sigma T} e^{j\omega T}$; therefore, points in RHP in the s-plane are mapped to points outside the unit circle in the Z-plane.

In summary, we have:

s-plane	Z-plane
$j\omega$ – axis	On the unit circle
LHP	Inside the unit circle
RHP	Outside the unit circle
All the points on the $j\omega$ – axis having value $\omega = \dfrac{\phi \cdot 2\pi n}{T}$	ϕ Therefore, z-transform is a multivalued transform

1.4. Transformation of Some Elementary Functions

1. The Unit Step Function

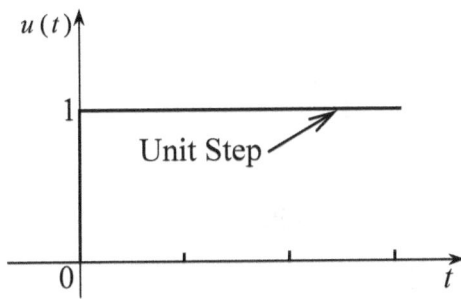

$$u(t) = \begin{cases} 0 & t < 0 \\ 1 & t \geq 0 \end{cases}$$

The sequence is

$\{u(k)\} = \{1, 1, 1, ...\}$, then

$$F(z) = 1 \cdot z^{0} + 1 \cdot z^{-1} + 1 \cdot z^{-2} + ... + 1 \cdot z^{-k} + ...$$

$$= \sum_{k=0}^{\infty} z^{-k} = \frac{1}{1 - \frac{1}{z}} = \frac{z}{z-1}$$

or $\boxed{\mathbb{Z}\{u(k)\} = \dfrac{z}{z-1}}$, with **ROC**: $z > 1$.

2. The Constant Function

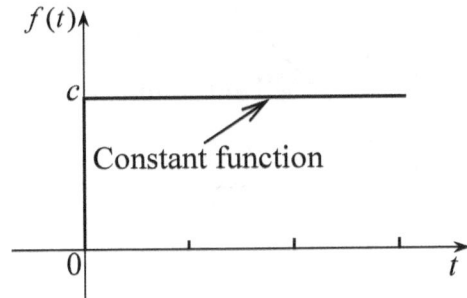

$f(t) = c$

The sequence is $f(k) = \{c, c, c, ...\}$, then

$$F(z) = c \cdot z^0 + c \cdot z^{-1} + c \cdot z^{-2} + \ldots + c \cdot z^{-k} + \ldots$$

$$= c \sum_{k=0}^{\infty} z^{-k} = \frac{c}{1 - \frac{1}{z}} = \frac{cz}{z-1}$$

or $\boxed{\mathbb{Z}\{c\} = \dfrac{cz}{z-1}}$, with **ROC**: $z > 1$.

3. The Unit Impulse Function

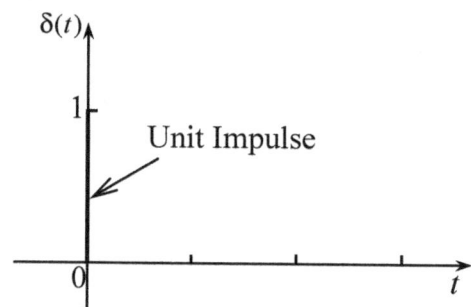

$f(t) = \delta(t)$

The sequence is $\delta(k) = \{1, 0, 0, \ldots\}$, then $F(z) = 1.z^0 = 1$.

or $\boxed{\mathbb{Z}\{\delta(k)\} = 1}$, with ROC = the entire complex plane.

4. The Unit Ramp Function

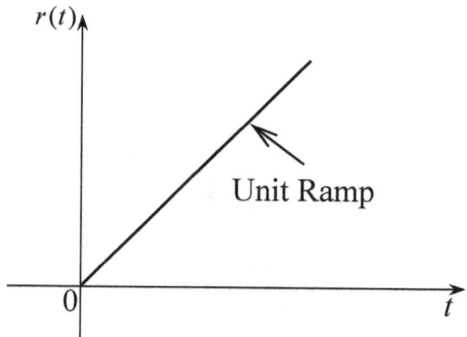

$r(t) = t, \quad t > 0$

The sequence is $r(k) = \{0, T, 2T, \ldots\}$,

then

$$F(z) = 0 \cdot z^0 + T \cdot z^{-1} + 2T \cdot z^{-2} + \ldots + kT \cdot z^{-k} + \ldots = \sum_{k=0}^{\infty} kT \cdot z^{-k}$$

$$\frac{zF(z)}{T} = 1 + 2z^{-1} + 3z^{-2} + \ldots = \frac{1}{(1-z^{-1})^2} = \frac{z^2}{(z-1)^2}$$

or $\boxed{\mathbb{Z}\{r(k)\} = \dfrac{Tz}{(z-1)^2}}$

5. The Exponential Function

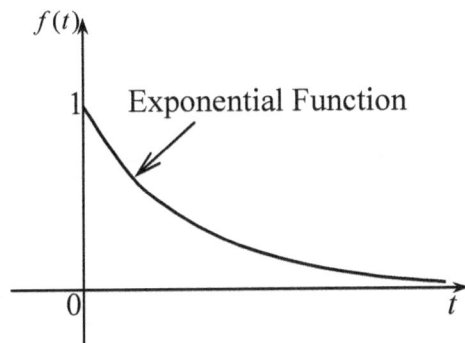

$$f(t) = e^{-at}, \quad t > 0$$

The sequence is $f(k) = \{1, e^{-aT}, e^{-2aT}, \ldots\}$, then

$$F(z) = 1 \cdot z^0 + e^{-aT} \cdot z^{-1} + e^{-2aT} \cdot z^{-2} + \ldots + e^{-k\,aT} \cdot z^{-k} + \ldots$$

$$= \sum_{k=0}^{\infty} e^{-k\,aT} \cdot z^{-k}$$

$$= 1 + (e^{aT} z)^{-1} + (e^{aT} z)^{-2} + \ldots$$

$$= \frac{1}{(1 - e^{-aT} z^{-1})} = \frac{z}{(z - e^{-aT})}$$

or $\boxed{\mathbb{Z}\{f(k)\} = \dfrac{z}{(z - e^{-aT})}}$, with **ROC**: $z > e^{aT}$.

6. Trigonometric Functions

Consider the function: $f(k) = e^{j\omega k}u(k)$. This function is an exponential function with a complex exponent. It is chosen because we know from the complex analysis theories that this function can be converted to trigonometric functions easily as we will see in our proof.

From the basic definition:

$$\mathbb{Z}\{f(k)\} = \sum_{k=0}^{\infty} e^{j\omega k}z^{-k} = \frac{1}{1-e^{j\omega}z^{-1}}.$$

Then $\mathbb{Z}\{e^{j\omega k}u(k)\} = \dfrac{1}{1-e^{j\omega}z^{-1}}.$

And we know that $e^{j\theta} = \cos\theta + j\sin\theta$.

Considering one-sided transform:

$$\mathbb{Z}\{\cos\omega k + j\sin\omega k\} = \frac{1}{1-\cos\omega\cdot z^{-1} - j\sin\omega\cdot z^{-1}}$$

Multiplying by the complex conjugate of the denominator:

$$\mathbb{Z}\{\cos\omega k + j\sin\omega k\} = \frac{1}{1-\cos\omega\cdot z^{-1} - j\sin\omega\cdot z^{-1}}$$
$$\times\frac{1-\cos\omega\cdot z^{-1} + j\sin\omega\cdot z^{-1}}{1-\cos\omega\cdot z^{-1} + j\sin\omega\cdot z^{-1}}$$

Simplifying, we get

$$\mathbb{Z}\{\cos\omega k + j\sin\omega k\}$$
$$= \frac{1-\cos\omega\cdot z^{-1} + j\sin\omega\cdot z^{-1}}{1-2\cos\omega\cdot z^{-1} + \cos^2\omega\cdot z^{-2} + \sin^2\omega\cdot z^{-2}}$$

Equating the real parts in both sides:

$$\mathbb{Z}\{\cos \omega k\} = \frac{1-\cos \omega \cdot z^{-1}}{1-2\cos \omega \cdot z^{-1}+z^{-2}}$$

Equating the imaginary parts in both sides:

$$\mathbb{Z}\{\sin \omega k\} = \frac{\sin \omega \cdot z^{-1}}{1-2\cos \omega \cdot z^{-1}+z^{-2}}$$

7. Hyperbolic Functions

Consider the function: $f(k) = e^{ak}u(k)$.
From the basic definition:

$$\mathbb{Z}\{f(k)\} = \sum_{k=0}^{\infty} e^{ak}z^{-k} = \frac{1}{1-e^{a}z^{-1}}.$$

Consider the function:

$$f(k) = e^{-ak}u(k)$$

From the basic definition:

$$\mathbb{Z}\{f(k)\} = \sum_{k=0}^{\infty} e^{-ak}z^{-k} = \frac{1}{1-e^{-a}z^{-1}}.$$

The chosen functions are exponential functions with a real exponent. They are chosen because we know that these functions can be converted to hyperbolic functions easily as we will see in our proof. Now using the definition of hyperbolic functions in terms of exponential function:

$$\mathbb{Z}\{\cosh ak\} = \frac{1}{2}\mathbb{Z}\{e^{ak}\} + \frac{1}{2}\mathbb{Z}\{e^{-ak}\}.$$

Using Z-transform of exponential functions:

$$\mathbb{Z}\{\cosh ak\} = \frac{\frac{1}{2}}{1-e^{a}z^{-1}} + \frac{\frac{1}{2}}{1-e^{-a}z^{-1}}$$

$$= \frac{\frac{1}{2}-\frac{1}{2}e^{-a}z^{-1}+\frac{1}{2}-\frac{1}{2}e^{a}z^{-1}}{(1-e^{a}z^{-1})(1-e^{-a}z^{-1})}$$

$$= \frac{1 - \frac{1}{2}(e^a z^{-1} + e^{-a} z^{-1})}{1 - e^a z^{-1} - e^{-a} z^{-1} + z^{-2}}$$

Converting the exponential functions to hyperbolic functions:

$$\boxed{\mathbb{Z}\{\cosh ak\} = \frac{1 - \cosh a \cdot z^{-1}}{1 - 2\cosh a \cdot z^{-1} + z^{-2}}}$$

$$\mathbb{Z}\{\sinh ak\} = \frac{1}{2}\mathbb{Z}\{e^{ak}\} - \frac{1}{2}\mathbb{Z}\{e^{-ak}\}$$

Using Z-transform of exponential functions:

$$\mathbb{Z}\{\sinh ak\} = \frac{\frac{1}{2}}{1 - e^a z^{-1}} - \frac{\frac{1}{2}}{1 - e^{-a} z^{-1}}$$

Simplifying:

$$\mathbb{Z}\{\sinh ak\} = \frac{\frac{1}{2} - \frac{1}{2}e^{-a} z^{-1} - \frac{1}{2} + \frac{1}{2}e^a z^{-1}}{(1 - e^a z^{-1})(1 - e^{-a} z^{-1})}$$

$$= \frac{\frac{1}{2}(e^a z^{-1} - e^{-a} z^{-1})}{1 - e^a z^{-1} - e^{-a} z^{-1} + z^{-2}}$$

Converting the exponential functions to hyperbolic functions:

$$\boxed{\mathbb{Z}\{\sinh ak\} = \frac{\sinh a \cdot z^{-1}}{1 - 2\cosh a \cdot z^{-1} + z^{-2}}}.$$

1.5. Properties of the Transform

From now on we will use the sampling interval $T = 1$ as to simplify the analysis. If any different sampling interval is used it will be clearly mentioned.

1.5.1. Linearity

Let $\mathbb{Z}\{f(k)\} = F(z)$, $\mathbb{Z}\{g(k)\} = G(z)$, and a, b be two arbitrary constants, then

$$\sum[af(k) + bg(k)]z^{-k} = a\sum f(k)z^{-k} + b\sum g(k)z^{-k} .$$

Thus $$\boxed{\mathbb{Z}\{af(k) + bg(k)\} = aF(z) + bG(z)}.$$

i.e. The Z-Transform is a linear transform.

Example: Using the linearity concept, find the Z-transform of the sampled function of $f(t) = \sin \omega t$. Compare your result with the one obtained before in the previous section.

Solution: The sampled function will be

$$f(k) = \sin \omega k .$$

Using Euler's expression $\quad \sin \omega k = \dfrac{1}{2j}[e^{j\omega k} - e^{-j\omega k}]$,

we obtain

$$\mathbb{Z}\{\sin \omega k\} = \frac{1}{2j}\left[\frac{z}{z - e^{j\omega}} - \frac{z}{z - e^{-j\omega}}\right]$$

$$= \frac{1}{2j}\left[\frac{z(e^{j\omega} - e^{-j\omega})}{z^2 - z(e^{j\omega} + e^{-j\omega}) + 1}\right] = \frac{z \sin \omega}{z^2 - 2z \cos \omega + 1} .$$

$$\boxed{\mathbb{Z}\{\sin \omega k\} = \frac{\sin \omega \cdot z^{-1}}{1 - 2\cos \omega \cdot z^{-1} + z^{-2}}}$$

which is the same result obtained before in the previous section.

1.5.2. Change of Scale

If $\mathbb{Z}\{f(k)\} = F(z)$, then $\boxed{\mathbb{Z}\{a^{-k}f(k)\} = F(az).}$

Proof: From the definition of Z-transform, we have,

$$\mathbb{Z}\{a^{-k}f(k)\} = \sum_{k=0}^{\infty} a^{-k}f(k)z^{-k} = \sum_{k=0}^{\infty} f(k)(az)^{-1} = F(az).$$

1.5.3. Complex Translation

If $\mathbb{Z}\{f(k)\} = F(z)$, then $\boxed{\mathbb{Z}\{e^{-ak}f(k)\} = F(e^{a}z).}$

Proof: From the definition of Z-transform, we have,

$$\mathbb{Z}\{e^{-ak}f(k)\} = \sum_{k=0}^{\infty} e^{-ak}f(k)z^{-k} = \sum_{k=0}^{\infty} f(k)(e^{a}z)^{-k} = F(e^{a}z).$$

Example: Find the Z-transform of the sampled function of $f(t) = te^{-at}$.

Solution: The sampled function will be $f(k) = ke^{-ak}$. We know that

$$\mathbb{Z}\{k\} = \frac{z}{(z-1)^2}, \quad \text{then}$$

$$\mathbb{Z}\{ke^{-ak}\} = \frac{ze^{a}}{(ze^{a}-1)^2} = \frac{e^{-a}z}{(z-e^{-a})^2}.$$

Example: Find the Z-transform of the sampled function of $f(t) = e^{-at}\sin \omega t$.

Solution: The sampled function will be $f(k) = e^{-ak}\sin \omega k$.

We know that $\mathbb{Z}\{\sin \omega k\} = \dfrac{z \sin \omega}{z^2 - 2z \cos \omega + 1}$, then,

$$\mathbb{Z}\{e^{-ak}\sin \omega k\} = \frac{(ze^{a})\sin \omega}{(ze^{a})^{2} - 2(ze^{a})\cos \omega + 1}$$

$$= \frac{ze^{-a}\sin \omega}{z^{2} - 2ze^{-a}\cos \omega + e^{-2a}}.$$

1.5.4. Multiplication by k

If $\mathbb{Z}\{f(k)\} = F(z)$, then $\boxed{\mathbb{Z}\{kf(k)\} = -z\dfrac{d}{dz}F(z).}$

Proof: From the definition of Z-transform, we have,

$$\mathbb{Z}\{kf(k)\} = \sum_{k=0}^{\infty} kf(k)z^{-k} = -z\sum_{k=0}^{\infty} -kf(k)z^{-k-1}$$

$$= -z\frac{d}{dz}\sum_{k=0}^{\infty} f(k)z^{-k} = -z\frac{d}{dz}F(z).$$

Example: Deduce the Z-transform of the sampled function of $f(t) = t^{2}$ from the Z-transform of 1.

Solution: Here $f(k) = k^{2}$. Now, we know that $\mathbb{Z}\{1\} = \dfrac{z}{z-1}$, then

$$\mathbb{Z}\{k\} = -z\frac{d}{dz}\frac{z}{z-1} = \frac{z}{(z-1)^{2}}$$

$$\mathbb{Z}\{k^{2}\} = -z\frac{d}{dz}\frac{z}{(z-1)^{2}} = \frac{z(z+1)}{(z-1)^{3}}.$$

The general form of the previous theorem is

If $\mathbb{Z}\{f(k)\} = F(z)$, then $\boxed{\mathbb{Z}\{k^{n}f(k)\} = \left(-z\dfrac{d}{dz}\right)^{n}F(z).}$

1.5.5. First Shift Theorem

This theorem is concerned with shifts of a sampled sequence *to the left*. If $\mathbb{Z}\{f(k)\} = F(z)$, then

$$\mathbb{Z}\{f\,(k+n)\} = z^{\,n} F(z) - \{z^{\,n} f\,(0) + z^{\,n-1} f\,(1) + ... + zf\,(n-1)\}.$$

Proof: From the definition of Z-transform, we have,

$$F(z) = f(0)z^0 + f(1)z^{-1} + f(2)z^{-2} + ...,$$

Multiplying by z, we get

$$zF(z) = f(0)z + f(1)z^0 + f(2)z^{-1} + ...$$

$$= zf(0) + \sum_{k=0}^{\infty} f(k+1)z^{-k} = zf(0) + \mathbb{Z}\{f(k+1)\}$$

Therefore $\mathbb{Z}\{f(k+1)\} = z\,F(z) - z\,f(0)$

Similarly, if we multiply by z^2, we obtain

$$z^2 F(z) = f(0)z^2 + f(1)z^1 + f(2)z^0 + f(2)z^{-1}...$$

$$= z^2 f(0) + zf(1) + \sum_{k=0}^{\infty} f(k+2)z^{-k}$$

$$= z^2 f(0) + zf(1) + \mathbb{Z}\{f(k+2)\}$$

Therefore $\mathbb{Z}\{f(k+2)\} = z^2 F(z) - [z^2 f(0) + zf(1)]$.

If we continue in the same manner, we can deduce that

$$\mathbb{Z}\{f(k+n)\} = z^{\,n} F(z) - \{z^{\,n} f(0) + z^{\,n-1} f(1) + ... + zf(n-1)\}.$$

**Example**: Suppose that the sequence $f(k)$ is given by

$$f(k) = \{0, 0, 0, 1, 1, 1, 1, ...\},$$

then the sequence $f(k+1)$ will be

$$f(k+1) = \{0, 0, 1, 1, 1, 1, 1, ...\};$$

and the sequence $f(k+2)$ will be

$$f(k+2) = \{0, 1, 1, 1, 1, 1, 1, ...\}.$$

These sequences are shown in Figure 6 below.

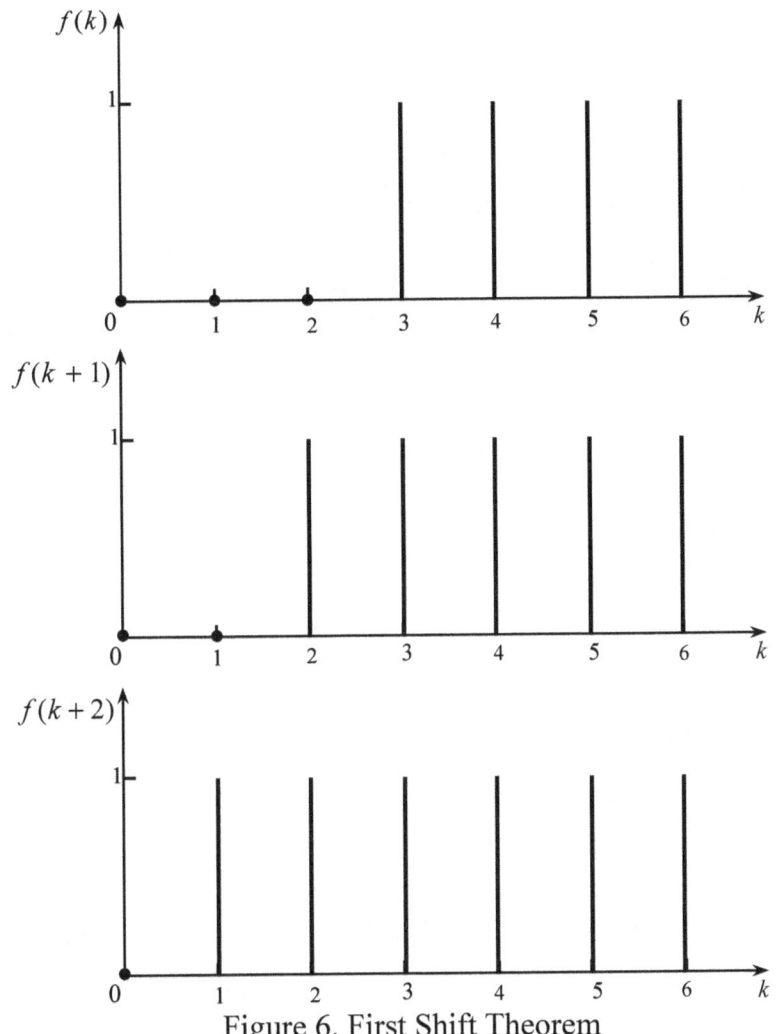

Figure 6. First Shift Theorem

1.5.6. Second Shift Theorem

This theorem is concerned with shifts of a sampled sequence *to the right*.

If $\mathbb{Z}\{f(k)\} = F(z)$, and if $f(t)u(t)$ is shifted to the right by n sample intervals, then the Z-transform of the shifted function is given by

$$\boxed{\mathbb{Z}\{f(k-n)u(k-n)\} = z^{-n}F(z).}$$

Proof: From the definition of Z-transform, we have,

$$\mathbb{Z}\{f(k-n)u(k-n)\} = \sum_{k=0}^{\infty} f(k-n)u(k-n)z^{-k}$$

But from the definition of the unit function

$$u(k-n) = \begin{cases} 0, & k < n \\ 1, & k > n \end{cases},$$

we obtain

$$\mathbb{Z}\{f(k-n)u(k-n)\} = \sum_{k=n}^{\infty} f(k-n)z^{-k}$$

$$= f(0)z^{-n} + f(1)z^{-n-1} + f(2)z^{-n-2} + \dots$$

$$= z^{-n}[f(0)z^{0} + f(1)z^{-1} + f(2)z^{-2} \dots$$

$$= z^{-n}F(z).$$

Example: Suppose that the function $f(t) = tu(t)$ is sampled at a sampling interval of one. Then the sampled function is $ku(k)$ and the sampling sequence is $\{0, 1, 2, 3, \dots\}$. When it is shifted to the right it gives the shifted function $(t-1)u(t-1)$ with sampled function $(k-1)u(k-1)$ whose sequence is $\{0, 0, 1, 2, 3, \dots\}$. This is shown in Figure 7.

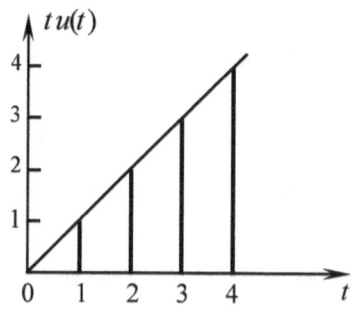 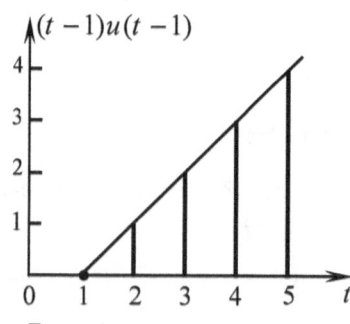

Figure 7.

Example: From the Z-transform of the sampled function of $tu(t)$ obtain the transform of the sampled function of $(t-1)u(t-1)$ when sampled at (i) $T = 1$ second and (ii) $T = 0.5$ second.

Solution: We know that $\mathbb{Z}\{ku(k)\} = \dfrac{zT}{(z-1)^2}$.

Using the Second Shift Theorem, we get

(i) At $T = 1$ second :

$$\mathbb{Z}\{(k-1)u(k-1)\} = z^{-1}\cdot\frac{z}{(z-1)^2} = \frac{1}{(z-1)^2}.$$

(ii) At $T = 0.5$ second :

$$\mathbb{Z}\{(k-2)u(k-2)\} = z^{-2}\cdot\frac{0.5z}{(z-1)^2} = \frac{0.5z^{-1}}{(z-1)^2}.$$

Example: From the Z-transform of the sampled function of t^2 obtain the transform of the sampled function of $(t+1)^2$ when sampled at $T = 1$ second.

Solution: We know that $\mathbb{Z}\{k^2\} = \dfrac{z^2 + z}{(z-1)^3}$, then

$$\mathbb{Z}\{(k+1)^2\} = z \cdot \frac{z^2 + z}{(z-1)^3} - [zf(0)]$$

But $f(0) = 0$, then $\mathbb{Z}\{(k+1)^2\} = \dfrac{z^2(z+1)}{(z-1)^3}$.

Note on the Shift Theorems:

In the shift theorems discussed above, we considered the one-sided (unilateral) Z-transform. But now we ask whether these theorems will change if we consider the two-sided (bilateral) Z-transform case where the input functions extend from $-\infty$ to ∞ .

<u>Left Time shifting for the two-sided case</u>

If $\mathbb{Z}\{f(k)\} = F(z)$, then $\boxed{\mathbb{Z}\{f(k+n)\} = z^n F(z).}$

Proof: From the definition of Z-transform, we have,

$$\mathbb{Z}\{f(k+n)\} = \sum_{k=-\infty}^{\infty} f(k+n)z^{-k} =$$

$$z^n \sum_{k=-\infty}^{\infty} f(k+n)z^{-(k+n)}$$

Let $k + n = m$, then

$$\mathbb{Z}\{f(k+n)\} = z^n \sum_{m=-\infty}^{\infty} f(m)z^{-(m)} = z^n F(z).$$

<u>Right Time shifting for the two-sided case</u>

If $\mathbb{Z}\{f(k)\} = F(z)$, then $\boxed{\mathbb{Z}\{f(k-n)\} = z^{-n} F(z).}$

Proof: From the definition of Z-transform, we have,

$$\mathbb{Z}\{f(k-n)\} = \sum_{k=-\infty}^{\infty} f(k-n)z^{-k}$$

$$= z^{-n} \sum_{k=-\infty}^{\infty} f(k-n)z^{-(k-n)}$$

Let $k - n = m$, then

$$\mathbb{Z}\{f(k-n)\} = z^{-n} \sum_{m=-\infty}^{\infty} f(m)z^{-(m)} = z^{-n} F(z)$$

1.5.7. Initial Value Theorem

The initial value theorem gives the value that the function $f(k)$ has when $k = 0$. This theorem is

$$\boxed{f(0) = \lim_{z \to \infty} F(z)}$$

provided that the limit exists.

Proof: From the definition of the Z-transform of $f(k)$, we have

$$F(z) = \sum_{k=0}^{\infty} f(k)z^{-k} = f(0) + f(1)z^{-1} + f(2)z^{-2} + \dots$$

Letting $z \to \infty$, we obtain

$$\lim_{z \to \infty} F(z) = \lim_{z \to \infty} [f(0) + f(1)z^{-1} + f(2)z^{-2} + \dots]$$

Thus $f(0) = \lim_{z \to \infty} F(z)$.

Example: Find the initial values of the sequences that have the following Z-transforms:

(i) $F(z) = \dfrac{z}{z-1}$ (ii) $F(z) = \dfrac{z}{(z-1)^2}$.

Solution: (i) $f(0) = \lim_{z \to \infty} \dfrac{z}{z-1} = 1$.

(ii) $f(0) = \lim_{z \to \infty} \dfrac{z}{(z-1)^2} = 0$.

Example: Use the initial value theorem, show that:

$$\cos 0 = 1 \quad \text{and} \quad \sin 0 = 0$$

Solution: $\cos 0 = \lim_{z \to \infty} \dfrac{1 - \cos \omega \cdot z^{-1}}{1 - 2\cos \omega \cdot z^{-1} + z^{-2}} = 1/1 = 1$

$$\sin 0 = \lim_{z \to \infty} \frac{\sin \omega \cdot z^{-1}}{1 - 2\cos \omega \cdot z^{-1} + z^{-2}} = 0/1 = 0$$

Note that $\cos \omega$ and $\sin \omega$ are both constants.

1.5.8. Final Value Theorem

The final value theorem gives the value that would be reached eventually by the time function $f(k)$; *i.e.* the steady state value. This theorem is

$$\boxed{\lim_{k \to \infty} f(k) = \lim_{z \to 1} \left\{ \frac{z-1}{z} F(z) \right\}}$$

provided that $F(z)$ gives a convergent sequence.

Proof: From the definition of the Z-transform, we have

$$\mathbb{Z}\{f(k+1) - f(k)\} = \lim_{n \to \infty} \sum_{k=0}^{n} [f(k+1) - f(k)] z^{-k}$$

Using the shift property, we get

$$z F(z) - z f(0) - F(z) = \lim_{n \to \infty} \sum_{k=0}^{n} [f(k+1) - f(k)] z^{-k}$$

Letting $z \to 1$, we obtain

$$\lim_{z \to 1} [(z-1)F(z) - z f(0)] = \lim_{z \to 1} \lim_{n \to \infty} \sum_{k=0}^{n} [f(k+1) - f(k)] z^{-k}$$

Interchanging the order of the limits, we get

$$\lim_{z \to 1} [(z-1)F(z)] = f(0) + \lim_{n \to \infty} \sum_{k=0}^{n} [f(k+1) - f(k)]$$

If we now expand the summation in the right hand side, we can observe that all terms will cancel out except the last term where $k \to \infty$, thus $\lim_{k \to \infty} f(k) = \lim_{z \to 1} \left\{ \frac{z-1}{z} F(z) \right\}$.

Example: What are the final values of the sequences that have the following Z-transforms:

(i) $F(z) = \dfrac{2z}{(z-1)(z^2 - 0.42z + 0.21)}$.

(ii) $F(z) = \dfrac{0.5z}{(z-1)(z-0.2)(z-0.3)}$

(iii) $F(z) = \dfrac{z}{(z-0.1)(z-1)}$.

Solution: Applying the Final Value Theorem, we obtain

(i) $\displaystyle\lim_{k \to \infty} f(k) = \lim_{z \to 1}\left\{ \frac{z-1}{z} \cdot \frac{2z}{(z-1)(z^2 - 0.42z + 0.21\}} \right\}$

$$= \frac{2}{(1 - 0.42 + 0.21)} = 2.53$$

(ii) $\displaystyle\lim_{k \to \infty} f(k) = \lim_{z \to 1}\left\{ \frac{z-1}{z} \cdot \frac{0.5z}{(z-1)(z-0.2)(z-0.3\}} \right\}$

$$= \frac{0.5}{(0.8)(0.7)} = 0.89$$

(ii) $\displaystyle\lim_{k \to \infty} f(k) = \lim_{z \to 1}\left\{ \frac{z-1}{z} \cdot \frac{z}{(z-0.1)(z-1)} \right\}$

$$= \frac{1}{0.9} = 1.11 .$$

1.5.9. Partial Differentiation Theorem

If $\mathbb{Z}\{f(k)\} = F(z)$, then

$$\boxed{\mathbb{Z}\{\tfrac{\partial}{\partial a}[f(k,a)]\} = \tfrac{\partial}{\partial a}[F(z,a)]}$$

<u>*Proof:*</u>   From the definition of the Z-transform, we have

$$\mathbb{Z}\{\tfrac{\partial}{\partial a}[f(k,a)]\} = \sum_{k=-\infty}^{\infty} \tfrac{\partial}{\partial a} f(k,a) z^{-k}$$

$$= \tfrac{\partial}{\partial a} \sum_{k=-\infty}^{\infty} f(k,a) z^{-k}$$

$$= \tfrac{\partial}{\partial a}[F(z,a)]$$

1.5.10. Convolution

The convolution of the functions $f(k)$ and $g(k)$ is defined by

$$f(k) * g(k) = \sum_{n=0}^{\infty} f(n)g(k-n)$$

1.5.10.1. The Convolution Theorem

If $\mathbb{Z}\{f(k)\} = F(z)$ with **ROC** $= R_1$ and $Z\{g(k)\} = G(z)$ with **ROC** $= R_2$, then $\boxed{\mathbb{Z}\{f(k)*g(k)\} = F(z)G(z)}$ with **ROC** containing $R_1 \cap R_2$.

Proof: From the definition of Z-transform, we have

$$\mathbb{Z}\{f(k)*g(k)\} = \sum_{k=0}^{\infty} [f(k)*g(k)]z^{-k}$$

$$= \sum_{k=0}^{\infty}\sum_{n=0}^{\infty} [f(n)g(k-n)]z^{-k}$$

$$= \sum_{n=0}^{\infty} f(n) \underbrace{\sum_{k=0}^{\infty} g(k-n)z^{-k}}_{z^{-n}G(z)}$$

$$= \left(\sum_{n=0}^{\infty} f(n)z^{-n}\right) G(z) = F(z)G(z).$$

This is the most important property of the Z-transform. It states that: *The convolution in the time domain corresponds to pointwise multiplication in the z-plane.* This has an implication in the design of recursive and nonrecursive filters among other applications.

Example: Find the Z-transform of the following function

$$f(k) = \begin{cases} a^k & \text{for } k \text{ even} \\ b^k & \text{for } k \text{ odd} \end{cases}$$

Solution: We have

$$\mathbb{Z}\{f(k)\} = \sum_{k=0}^{\infty} a^{2k} z^{-2k} + \sum_{k=0}^{\infty} b^{2k+1} z^{-(2k+1)}$$

$$= \sum_{k=0}^{\infty} \left(a^2 z^{-2}\right)^k + bz^{-1} \sum_{k=0}^{\infty} \left(b^2 z^{-2}\right)^k$$

Using the binomial theorem, we get

$$\mathbb{Z}\{f(k)\} = \frac{1}{1-a^2 z^{-2}} + \frac{bz^{-1}}{1-b^2 z^{-2}}$$

Example: Find the Z-transform of the following function

$$f(k) = \begin{cases} 1 & 0 \le k < 8 \\ 0 & k \ge 8 \end{cases}$$

Solution: <u>Method (1)</u>: From the basic definition:

$$\mathbb{Z}\{f(k)\} =$$

$$1 + z^{-1} + z^{-2} + z^{-3} + z^{-4} + z^{-5} + z^{-6} + z^{-7}$$

<u>Method (2)</u>: Write the function $f(k) = u(k) - u(k-8)$

$$\mathbb{Z}\{f(k)\} = \frac{1}{1-z^{-1}} - z^{-8}\frac{1}{1-z^{-1}} = \frac{1-z^{-8}}{1-z^{-1}}$$

We can prove that the two solutions are equal using factorization and simplifying of the second solution.

Example: Find the Z-transform of the following function:

$$f(k) = \frac{a^k}{k!}.$$

Solution: From the basic definition: $\mathbb{Z}\{f(k)\} = \displaystyle\sum_{k=0}^{\infty}\frac{a^k}{k!}z^{-k}$

Using the binomial theorem:

$$\mathbb{Z}\{f(k)\} = \sum_{k=0}^{\infty}\frac{\left(az^{-1}\right)^k}{k!} = e^{az^{-1}}.$$

Note: It is quite common to have to determine the Z-transform that corresponds to a particular Laplace transform. It is to be noted that it is not obtained by simply substituting z for s. A basic approach that can be used is to determine the inverse Laplace transform, *i.e.* find $f(t)$ then find the corresponding $f(k)$ after sampling, and finally take the Z-transform of the sampled signal. ***Table* (1)** gives the Laplace and Z-transform for some basic functions. Also ***Table* (2)** gives some of the properties of the Z-transform.

Example: Determine the Z-transform for the sampled functions for which:

$$(i)\ F(s) = \frac{1}{(s+a)(s+b)};\qquad (ii)\ F(s) = \frac{1}{(s^2+1)(s^2+4)}$$

Solution: (*i*) Using partial fraction decomposition, we have

$$F(s) = \frac{1}{b-a}\left[\frac{1}{s+a} - \frac{1}{s+b}\right]$$

Taking the inverse Laplace transform

$$f(t) = \frac{1}{b-a}\left[e^{-at} - e^{-bt}\right]$$

The corresponding sampling function with a sampling period T is

$$f^*(k) = \frac{1}{b-a}\left[e^{-akT} - e^{-bkT}\right]$$

Taking the Z-transform, we obtain

$$F(z) = \frac{1}{b-a}\left[\frac{z}{z-e^{-aT}} - \frac{z}{z-e^{-bT}}\right]$$

$$= \frac{1}{b-a}\left[\frac{z(e^{-aT} - e^{-bT})}{(z-e^{-aT})(z-e^{-bT})}\right]$$

(*ii*) Using partial fraction decomposition, we have

$$F(s) = \frac{1}{3}\left[\frac{1}{s^2+1} - \frac{1}{s^2+4}\right]$$

Taking the inverse Laplace transform

$$f(t) = \frac{1}{3}\left[\sin t - \frac{1}{2}\sin 2t\right]$$

The corresponding sampling function with a sampling period T is

$$f^*(k) = \frac{1}{3}\left[\sin kT - \frac{1}{2}\sin 2kT\right]$$

Taking the Z-transform, we obtain

$$F(z) = \frac{1}{3}\left[\frac{z\sin T}{z^2 - 2z\cos T + 1} - \frac{1/2\,z\sin 2T}{z^2 - 2z\cos 2T + 1}\right].$$

Table 1. Z-transform Table

$f(k)$	$F(z)$
$\delta(k)$	1
$\delta(k-n)$	z^{-n}
$u(k)$	$\dfrac{z}{z-1}$
$a^{k-1}u(k-1)$	$\dfrac{1}{z-a}$
$r(k) = kT$	$\dfrac{Tz}{(z-1)^2}$
$(kT)^2$	$\dfrac{T^2 z(z+1)}{(z-1)^3}$
$(kT)^3$	$\dfrac{T^2 z(z^2 + 4z + 1)}{(z-1)^4}$
e^{-akT}	$\dfrac{z}{z-e^{-aT}}$
$1 - e^{-akT}$	$\dfrac{z(1-e^{-aT})}{(z-1)(z-e^{-aT})}$
kTe^{-akT}	$\dfrac{T\,z\,e^{-aT}}{(z-e^{-aT})^2}$
$(1-akT)e^{-akT}$	$\dfrac{z\,[z - e^{-aT}(1-e^{-aT})]}{(z-e^{-aT})^2}$
$e^{-akT} - e^{-bkT}$	$\dfrac{z\,(e^{-aT} - e^{-bT})}{(z-e^{-aT})(z-e^{-bT})}$

c^k	$\dfrac{z}{z-c}$
kc^k	$\dfrac{cz}{(z-c)^2}$
$k^2 c^k$	$\dfrac{cz(z+c)}{(z-c)^3}$
$\sin \omega kT$	$\dfrac{z \sin \omega T}{z^2 - 2z \cos \omega T + 1}$
$\cos \omega kT$	$\dfrac{z(z - \cos \omega T)}{z^2 - 2z \cos \omega T + 1}$
$e^{-akT} \sin \omega kT$	$\dfrac{ze^{-aT} \sin \omega T}{z^2 - 2ze^{-aT} \cos \omega T + e^{-2aT}}$
$e^{-akT} \cos \omega kT$	$\dfrac{z(z - e^{-aT} \cos \omega T)}{z^2 - 2ze^{-aT} \cos \omega T + e^{-2aT}}$
$|a|^k \sin \omega k$	$\dfrac{z|a|\sin \omega}{z^2 - 2z|a|\cos \omega + |a|^2}$
$|a|^k \cos \omega k$	$\dfrac{z[z - |a|\sin \omega]}{z^2 - 2z|a|\cos \omega + |a|^2}$
$\sinh \omega kT$	$\dfrac{z \sinh \omega T}{z^2 - 2z \cosh \omega T + 1}$
$\cosh \omega kT$	$\dfrac{z(z - \cosh \omega T)}{z^2 - 2z \cosh \omega T + 1}$

Table 2. Properties of the Z-transform

Operation	$f(k)$	$F(z)$
Addition	$f_1(k) + f_2(k)$	$F_1(z) + F_2(z)$
Scalar Multiplication	$af(k)$	$aF(z)$
Right Shift	$f(k-n)$	$z^{-n}F(z)$
Left Shift	$f(k+n)$	$z^n F(z) - z^n \sum_{k=0}^{n-1} f(k)z^{-k}$
Multiplication by a^k	$a^k f(k)$	$F\left(\dfrac{z}{a}\right)$
Multiplication by k	$kf(k)$	$-z\dfrac{d}{dz}F(z)$
Time Convolution	$f_1(k) * f_2(k)$	$F_1(z)F_2(z)$
Frequency Convolution	$f_1(k)f_2(k)$	$\dfrac{1}{2\pi i}\oint F_1(u)F_2\left(z/u\right)u^{-1}du$
Initial Value	$f(0)$	$\lim_{z \to \infty} F(z)$
Final Value	$\lim_{k \to \infty} f(k)$	$\lim_{z \to 1}(z-1)F(z)$ poles of $(z-1)F(z)$ inside the unit circle

1.6. Exercises

1. Sketch the following sequences for $k = 0, 1, 2, \ldots$

   (a) $f(k) = (0.5)^k$ (b) $f(k) = 2\delta(k-1) + e^{-0.2k}\left[u(k-5) - u(k-8)\right]$

   (c) $f(k) = 3\cos(0.4k) + 4\sin(0.4k)$

2. Find the Z-transform of each of the following functions. Put your answer in the simplest form:

   (a)$f(k) = \delta(k-1) + 2\delta(k-2) + 3\delta(k-3)$

   $$\text{Ans}: z^{-1} + 2z^{-2} + 3z^{-3}$$

   (b)$f(k) = \left(\dfrac{1}{2}\right)^k u(k)$ $\text{Ans}: \dfrac{1}{1 - \dfrac{1}{2}z^{-1}}$ $ROC: |z| > \dfrac{1}{2}$

   (c)$f(k) = \left(\dfrac{1}{2}\right)^k$ *for* $k > 5$ $\text{Ans}: \dfrac{1}{64}\dfrac{z^{-6}}{1 - \dfrac{1}{2}z^{-1}}$ $ROC: |z| > \dfrac{1}{2}$

   $\quad\quad\quad = 0$ *for* $k \leq 5$

   (d)$f(k) = 3e^{-8k}u(k)$ $\text{Ans}: \dfrac{3z}{z - e^{-8}}$

   (e)$f(k) = (0.2)^k - 3(0.1)^k$ $\text{Ans}: \dfrac{z}{z - 0.2} - \dfrac{3z}{z - 0.1}$

   (f)$f(k) = (-1)^k$ *for* $k = 4, 5, 6, \ldots\ldots$ $\text{Ans}: \dfrac{1}{z^3(z+1)}$

   $\quad\quad\quad = 0$ otherwise

   (g)$f(k) = u(k-3)$ $\text{Ans}: z^{-3}/(1 - z^{-1})$

   (h)$f(k) = 0.1u(k) - 0.1\delta(k) - 0.1\delta(k-1)$ $\text{Ans}: \dfrac{0.1z^{-1}}{z-1}$

   (i)$f(k) = (1+k)\cdot u(k)$ $\text{Ans}: 1/(1-z^{-1})^2$ $ROC: |z| > 1$

$(j)f(k) = 0.34k\, u(k)$ $\qquad$ Ans: $\dfrac{0.34z}{(z-1)^2}$

$(k)f(k) = \left(a^k + a^{-k}\right)\cdot u(k), |a|>1$

$\qquad$ Ans: $\dfrac{2-\left(a+a^{-1}\right)z^{-1}}{1-\left(a^{-1}+a\right)z^{-1}+z^{-2}}$ $\quad$ ROC $:|z|>|a|$

$(l)f(k) = \left(\dfrac{1}{2}\right)^k \sin\left(\dfrac{\pi k}{2}\right)u(k)$ $\qquad$ Ans: $\dfrac{z^{-1}}{2(1+z^{-1}/4)}$

$(m)f(k) = \cos\left(2k+\dfrac{\pi}{6}\right)$ $\qquad$ Ans: $\dfrac{0.866z^2 - 0.0942z}{z^2 + 0.8323z + 1}$

$(n)f(k) = ke^{-0.2k}\sin(2k)$ $\qquad$ Ans: $\dfrac{0.7445z^3 - 0.499z}{\left(z^2 + 0.6814z - 0.6703\right)^2}$

$(o)f(k) = Ar^k\cos(\omega k+\phi)\cdot u(k), 0 < r < 1$

$\qquad$ Ans: $A\left\{\dfrac{\cos\phi - \cos(\omega-\phi)rz^{-1}}{1-2rz^{-1}\cos\omega + r^2 z^{-2}}\right\}$ ROC $:|z|>r$

$(p)f(k) = a^{|k|},\ |a|<1$ $\quad$ Ans: $\dfrac{\left(1-a^2\right)z^{-1}}{-a+\left(1+a^2\right)z^{-1}-az^{-2}}$ $\quad$ ROC$:|a|<|z|<\left|\dfrac{1}{a}\right|$

$(q)f(k) = \dfrac{1}{k!}$ $\qquad$ Ans: $(e)^{z^{-1}}$

$(r)f(k) = k\left(\delta(k-4) + u(k)\right)$ $\qquad$ Ans: $4z^{-4} + \dfrac{z^{-1}}{\left(1-z^{-1}\right)^2}$

3. Determine the Z-transform of the analog signal $x(t) = 10e^{-2t}u(t)$ after it goes through an ADC with $T = 0.01$s.   Ans: $\dfrac{10z}{z - e^{-0.02}}$

4. Determine the Z-transform of the analog signal:

   $x(t) = 21\sin(5t)u(t)$ after it goes through an ADC with $T = 0.01$s.

   $$\text{Ans}: \frac{21z\ \sin(0.05)}{z^2 - 2\cos(0.05) + 1}$$

5. Determine the Z-transform of the analog signal $x(t) = 3t\,u(t)$ after it goes through an ADC with $T = 0.03$s.   Ans: $\dfrac{0.09z}{(z-1)^2}$

6. Determine the Z-transform of the analog signal with the Laplace transform $X(s) = \dfrac{5}{s+10}$ after being sampled with $T = 0.025$s.

   $$\text{Ans}: \frac{1.4z}{z - e^{0.01}}$$

7. If $f(t) = 2 - 2\cos(2\pi t)$ is sampled every T seconds. Compute the Z-transform of the sampled sequence $f(k)$ for $T = 0.1$s.

   $$\text{Ans}: \frac{0.38z\,(z+1)}{(z-1)(z^2 - 1.618z + 1)}.$$

Refaat El Attar

Chapter Two

The Inverse ℤ -Transform

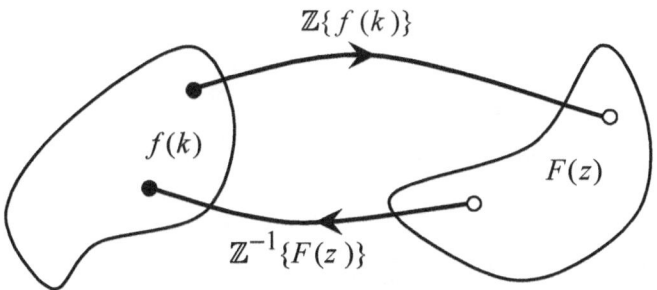

Chapter 2

The Inverse $\mathbb{Z}$ -Transform

2.1. The Inverse $\mathbb{Z}$ -Transform

The sequence of samples represented by a Z-transform, *i.e.* the inverse Z-transform defined as

$$\mathbb{Z}^{-1}\{F(z)\} = f(k)$$

can be obtained in a number of ways:

1. Partial Fractions

2. Long Division

3. The Residue Method (Inversion Formula Method)

2.2. The Partial Fraction Method

First of all, we are given the function $F(z)$ as a fraction and we are going to apply a partial fraction expansion for it. All methods of finding partial fractions may be used. These methods include equating coefficients, Heaviside method and finding partial fractions by inspection.

The Z-transform of the function is decomposed to the sum of more manageable simple functions using the partial fraction expansion. These simple functions can then be looked up in the Z-transform table. Because all the standard transforms have the term z in their numerators, we generally work with $F(z)/z$ instead of $F(z)$.

Note that if the denominator is a polynomial of the first degree, we expect the time domain function to be an exponential function and if it is a polynomial of the second degree, we expect the time domain function to be a trigonometric or hyperbolic function.

Here are some useful transform pairs for inversion by partial fractions.

$F(z);\ |z|>|a|$	$f(k);\ k>0$
$\dfrac{z}{z-a}$	a^k
$\dfrac{z^2}{(z-a)^2}$	$(k+1)a^k$
$\dfrac{z^3}{(z-a)^3}$	$\dfrac{(k+1)(k+2)}{2!}a^k$
$\dfrac{z^n}{(z-a)^n}$	$\dfrac{(k+1)(k+2)....(k+n-1)}{(n-1)!}a^k\ ;n\geq 2$
$\dfrac{z}{z-a}$	$-a^k$
$\dfrac{z^2}{(z-a)^2}$	$-(k+1)a^k$
$\dfrac{z^3}{(z-a)^3}$	$-\dfrac{(k+1)(k+2)}{2!}a^k$
$\dfrac{z^n}{(z-a)^n}$	$-\dfrac{(k+1)(k+2)....(k+n-1)}{(n-1)!}a^k\ ;n\geq 2$

Example: Use the partial fraction method to determine the inverse Z-transform for $F(z)=\dfrac{0.33z}{(z-e^{-0.4})(z-1)}$.

Solution: First, divide by z, we get $\dfrac{F(z)}{z}=\dfrac{0.33}{(z-e^{-0.4})(z-1)}$

Using partial fraction decomposition for the right-hand-side, we obtain

$$\dfrac{F(z)}{z}=\dfrac{A}{z-e^{-0.4}}+\dfrac{B}{z-1}$$

By inspection we can see that $A=1$ and $B=-1$. Then

$$F(z)=\dfrac{z}{z-e^{-0.4}}-\dfrac{z}{z-1}$$

From Table 1, we can deduce the inverse transform

$$f(k) = e^{-0.4k} - 1$$

The sequence is thus

$$f(k) = \{0, -0.33, -0.55, -0.70, ...\}$$

If the sampling period is $T = 1$ second, we could have written

$$F(z) = \frac{z}{z - e^{-0.4T}} - \frac{z}{z - 1},$$

then $f(t) = e^{-0.4t} u(t) - u(t)$.

On the other hand, if the sampling period is $T = 0.5$ second, we could have written

$$F(z) = \frac{z}{z - e^{-0.8T}} - \frac{z}{z - 1},$$

then, $f(t) = e^{-0.8t} u(t) - u(t)$.

Example: Use the partial fraction method to find the sampled function

for $\quad F(z) = \dfrac{z}{(z-3)(z-2)^2} \quad$ when $T = 1\,s$, $T = 0.5\,s$ and

$T = 0.2\,s$.

Solution: First, divide by z, we get

$$\frac{F(z)}{z} = \frac{1}{(z-3)(z-2)^2}.$$

Using partial fraction decomposition for the right-hand-side, we obtain

$$\frac{F(z)}{z} = \frac{A}{z-3} + \frac{B}{(z-2)^2} + \frac{C}{z-2}$$

We get

$$A = 1, \ B = -1 \text{ and } C = -1.$$

Then

$$\frac{F(z)}{z} = \frac{1}{z-3} - \frac{1}{(z-2)^2} - \frac{1}{z-2}, \text{ or}$$

$$F(z) = \frac{z}{z-3} - \frac{z}{(z-2)^2} - \frac{z}{z-2}$$

From **Table 1**, we can deduce the inverse transform

$$f(k) = 3^k - k(2)^k - 2^k$$

For $T = 1\,s$, $F(z) = \dfrac{z}{z-3^T} - \dfrac{1}{2}\dfrac{z.2^T}{(z-2^T)^2} - \dfrac{z}{z-2^T}$,

and

$$f(t) = 3^t - \frac{1}{2}t.2^t - 2^t.$$

For $T = 0.5\,s$,

$$F(z) = \frac{z}{z-3^{2T}} - \frac{z.2^{2T}(2T)}{2^{2T}(z-2^{2T})^2} - \frac{z}{z-2^{2T}}$$

$$= \frac{z}{z-3^{2T}} - \frac{T2^{2T}z}{(z-2^{2T})^2} - \frac{z}{z-2^{2T}}$$

and $f(t) = 3^{2t} - t.2^{2t} - 2^{2t}.$

For $T = 0.2\,s$,

$$F(z) = \frac{z}{z-3^{5T}} - \frac{z.2^{5T}(5T)}{2^{5T}(z-2^{5T})^2} - \frac{z}{z-2^{5T}}$$

$$= \frac{z}{z-3^{5T}} - \frac{5}{2}\frac{T2^{5T}z}{(z-2^{5T})^2} - \frac{z}{z-2^{5T}}$$

and $f(t) = 3^{5t} - \dfrac{5}{2}t.2^{5t} - 2^{5t}.$

The sequence of pulses is the same in each case but the function that generates them will depend on the sampling period used. The sequence is $\{0, 0, 1, ...\}$.

Example: Use the partial fraction method to determine the inverse Z-transform for $F(z) = \dfrac{6z^2 - 10z + 2}{z^2 - 3z + 2}$.

Solution: First, divide by z and factor the denominator, we get

$$\frac{F(z)}{z} = \frac{6z^2 - 10z + 2}{z(z-1)(z-2)}$$

Using partial fraction decomposition for the right-hand-side, we obtain

$$\frac{F(z)}{z} = \frac{A}{z} + \frac{B}{z-1} + \frac{C}{z-2}$$

The constants can be evaluated giving $A = 1$, $B = 2$ and $C = 3$. Then

$$\frac{F(z)}{z} = \frac{1}{z} + \frac{2}{z-1} + \frac{3}{z-2}, \text{ or}$$

$$F(z) = 1 + \frac{2z}{z-1} + \frac{3z}{z-2}$$

From Table 1, we can deduce the inverse transform

$$f(k) = \delta(k) + 2(1)^k + 3(2)^k$$

Example: Use the partial fraction method to determine the inverse Z-transform for $F(z) = \dfrac{7z}{10z^2 + 3z - 1}$. $\dfrac{1}{5} < z < \dfrac{1}{2}$

Solution: Note that the region of convergence here lies between two circles in the complex plane which means that $f(k)$ is a two-sided signal having an expression for positive values of k and a different one for negative values of k.

$$F(z) = \frac{7z}{10z^2 + 3z - 1} = \frac{\dfrac{7}{10}z}{z^2 + \dfrac{3}{10}z - \dfrac{1}{10}}$$

Factorizing the denominator: $F(z) = z\dfrac{\dfrac{7}{10}}{\left(z + \dfrac{1}{2}\right)\left(z - \dfrac{1}{5}\right)}$

Performing partial fraction expansion:

$$F(z)= \frac{7}{10}z\left(\frac{\frac{10}{7}}{\left(z-\frac{1}{5}\right)} - \frac{\frac{10}{7}}{\left(z+\frac{1}{2}\right)}\right) = \frac{z}{\left(z-\frac{1}{5}\right)} - \frac{z}{\left(z+\frac{1}{2}\right)}$$

It is clear that the first fraction corresponds to the positive part of the function as it converges outside the circle while the second fraction corresponds to the negative part of the function as it converges inside the circle.

$$\text{So, } f(k) = \begin{cases} (-\frac{1}{2})^k & k < 0 \\ (\frac{1}{5})^k & k \geq 0 \end{cases}$$

Example: Use the partial fraction method to determine the inverse Z-transform for $F(z) = \dfrac{z^3 - z^2 + z - \dfrac{1}{16}}{z^3 - \dfrac{5}{4}z^2 + \dfrac{1}{2}z - \dfrac{1}{16}}$

Solution: $F(z) = \dfrac{z^3 - z^2 + z - \dfrac{1}{16}}{z^3 - \dfrac{5}{4}z^2 + \dfrac{1}{2}z - \dfrac{1}{16}}$

Performing long division:

$$F(z) = 1 + \frac{-\dfrac{1}{4}z^2 + \dfrac{1}{2}z}{z^3 - \dfrac{5}{4}z^2 + \dfrac{1}{2}z - \dfrac{1}{16}}$$

Factorizing the numerator:

$$F(z) = 1 + z\frac{\dfrac{1}{4}(z+2)}{z^3 - \dfrac{5}{4}z^2 + \dfrac{1}{2}z - \dfrac{1}{16}}$$

Performing Partial fraction expansion:

$$F(z) = 1 + z\left(\frac{-9}{z - \dfrac{1}{2}} + \frac{\dfrac{5}{2}}{\left(z - \dfrac{1}{2}\right)^2} + \frac{9}{z - \dfrac{1}{4}} \right)$$

$$F(z) = 1 - 9\frac{z}{z - \dfrac{1}{2}} + 5\frac{z/2}{\left(z - \dfrac{1}{2}\right)^2} + 9\frac{z}{z - \dfrac{1}{4}}$$

Note that the denominator of the second fraction is squared, which indicates that it is obtained by differentiation of a certain function in Z- domain, which corresponds to multiplication, by k in the time domain.

So,

$$f(k) = \delta(k) - 9\left(\frac{1}{2}\right)^k u(k) + 5k\left(\frac{1}{2}\right)^k u(k) + 9\left(\frac{1}{4}\right)^k u(k)$$

Example: Use the partial fraction method to determine the inverse Z-transform for:

$$F(z) = \frac{z^2 - \dfrac{1}{4}z}{z^2 - \dfrac{3}{4}z + \dfrac{1}{8}} + \frac{2z^3}{(z^2 + 1)\cdot\left(z^2 - \dfrac{3}{4}z + \dfrac{1}{8}\right)}$$

Solution: $F(z) = \dfrac{z^2 - \dfrac{1}{4}z}{z^2 - \dfrac{3}{4}z + \dfrac{1}{8}} + \dfrac{2z^3}{(z^2 + 1)\cdot\left(z^2 - \dfrac{3}{4}z + \dfrac{1}{8}\right)}$

Performing Partial fraction expansion:

$$F(z) = \frac{z}{z - \dfrac{1}{2}} + \frac{8}{5}\cdot\frac{z}{z - \dfrac{1}{2}} - \frac{8}{17}\cdot\frac{z}{z - \dfrac{1}{4}} + \frac{\dfrac{112}{85}z^2 - \dfrac{96}{85}z}{z^2 + 1}$$

Simplifying:

$$F(z) = \frac{13}{5} \cdot \frac{z}{z - \frac{1}{2}} - \frac{8}{17} \cdot \frac{z}{z - \frac{1}{4}} + \frac{112}{85} \cdot \frac{z^2}{z^2 + 1} - \frac{96}{85} \cdot \frac{z}{z^2 + 1}$$

Note that the second-degree denominator will lead to a function containing sines and cosines in time domain.

$$f(k) = \frac{13}{5}\left(\frac{1}{2}\right)^k - \frac{8}{17}\left(\frac{1}{4}\right)^k + \frac{112}{85}\sin\frac{\pi}{2}k - \frac{96}{85}\cos\frac{\pi}{2}k; \quad k \geq 0.$$

2.3. The Long Division Method

We can use long division to find the inverse transform. This will give us the time sequence, but the technique fails to provide the function that generates this sequence.

Example: Use Long division to determine the time sequence corresponding to the image function $F(z) = \dfrac{2z}{z^2 + z + 1}$.

Solution: The procedure is to divide the numerator by the denominator using long division

$$
\begin{array}{r}
2z^{-1} - 2z^{-2} + 2z^{-4} - 2z^{-5} + \ldots \\
\hline
z^2 + z + 1 \;\big)\; 2z \qquad\qquad\qquad\qquad\qquad \\
2z + 2 + 2z^{-1} \qquad\qquad\qquad \\
\hline
-2 - 2z^{-1} \qquad\qquad\qquad \\
-2 - 2z^{-1} - 2z^{-2} \qquad\qquad \\
\hline
-2z^{-2} \qquad\qquad \\
-2z^{-2} + 2z^{-3} + 2z^{-4} \qquad \\
\hline
-2z^{-3} - 2z^{-4} \qquad \\
-2z^{-3} - 2z^{-4} - 2z^{-5} \\
\hline
+2z^{-5}
\end{array}
$$

Thus, we can represent $F(z)$ by the series

$$F(z) = 2z^{-1} - 2z^{-2} + 2z^{-4} - 2z^{-5} + \ldots$$

This is the transform of a sequence of impulses. Comparing with

$$F(z) = f(0)z^0 + f(1)z^{-1} + f(2)z^{-2} + f(3)z^{-3} + \ldots$$

we get the time sequence

$$f(k) = \{0, 2, -2, 0, 2, -2, \ldots\}.$$

Example: Find the inverse Z-transform of $F(z) = \dfrac{4z}{(z-1)^2}$.

Solution: This inverse can be found directly from **Table 1** as

$$f(k) = 4k \text{ or } f(k) = \{0, 4, 8, 12, 16, 20, ...\} .$$

But for the sake of illustrating the long division procedure, we have

$$
\begin{array}{r}
4z^{-1} + 8z^{-2} + 12z^{-3} + 16z^{-4} + ... \\
\hline
z^2 - 2z + 1 \, \big)\, 4z \\
4z - 8z^0 + 4z^{-1} \\
\hline
+8z^0 - 4z^{-1} \\
+8z^0 - 16z^{-1} + 8z^{-2} \\
\hline
12z^{-1} - 8z^{-2} \\
12z^{-1} - 24z^{-2} + 12z^{-3} \\
\hline
16z^{-2} - 12z^{-3} \\
16z^{-2} - 32z^{-3} + 16z^{-4} \\
\hline
20z^{-3} - 16z^{-4}
\end{array}
$$

Thus, we can represent $F(z)$ by the series

$$F(z) = 0z^0 + 4z^{-1} + 8z^{-2} + 12z^{-3} + 16z^{-4} + ...$$

This is the transform of a sequence of impulses. Comparing with

$$F(z) = f(0)z^0 + f(1)z^{-1} + f(2)z^{-2} + f(3)z^{-3} + f(4)z^{-4} + ...$$

we get the same time sequence

$$f(k) = \{0, 4, 8, 12, 16, ...\} .$$

Example: Find the inverse Z-transform of $F(z) = \dfrac{z}{z^2 - 2z + 4}$.

Solution: Using the long division procedure, we have

$$z^{-1} + 2z^{-2} - 8z^{-4} + \ldots$$

$$z^2 - 2z + 4 \enclose{longdiv}{z}$$

$$z - 2z^0 + 4z^{-1}$$

$$+2z^0 - 4z^{-1}$$

$$+2z^0 - 4z^{-1} + 8z^{-2}$$

$$-8z^{-2}$$

$$-8z^{-2} + 16z^{-3} - 32z^{-4}$$

$$-16z^{-3} + 32z^{-4}$$

Thus, we can represent $F(z)$ by the series

$$F(z) = 0z^0 + z^{-1} + 2z^{-2} - 4z^{-4} + \ldots$$

This is the transform of a sequence of impulses. Comparing with

$$F(z) = f(0)z^0 + f(1)z^{-1} + f(2)z^{-2} + f(3)z^{-3} + f(4)z^{-4} + \ldots \text{w}$$
e get the time sequence $f(k) = \{0, 1, 2, 0, -8, \ldots\}$.

Also, we could have put $F(z)$ in the form

$$F(z) = \frac{z}{z^2 - 2z\,(\cos \pi/3) + 2^2} = \frac{1}{2\sin \pi/3}\left[\frac{2z \sin \pi/3}{z^2 - 2z\,(\cos \pi/3) + 2^2}\right]$$

and from *Table* 1, the inverse is

$$f(k) = \frac{1}{\sqrt{3}}.2^k \sin\left[\frac{\pi}{3}k\right].$$

Note: The previous method is applied on a small number of problems because it fails to provide the function that generates the resulting time sequence. However, in some cases where we are interested in only a few sample values of $f(k)$, we may apply this method. For example, if $f(k)$ represents a system response and it decreases very rapidly to zero, we can for all practical purposes, evaluate just the first few values of $f(k)$ and assume that the rest are zero.

2.4. Residue Method (Inversion Formula Method)

From the definition of Z-transform

$$F(z) = \sum_{k=0}^{\infty} f(k)z^{-k}$$

If we multiply both sides by z^{n-1} and integrate along a closed contour C with respect to z, we get

$$\oint_C z^{n-1}F(z)dz = \oint_C z^{n-1}\left(\sum_{k=0}^{\infty} f(k)z^{-k}\right)dz$$

Interchanging the order of integration and summation, we obtain

$$\oint_C z^{n-1}F(z)dz = \sum_{k=0}^{\infty} f(k)\oint_C z^{(-k+n-1)}dz$$

From the theory of complex variables, we have

$$\oint_C z^m dz = \begin{cases} 2\pi i, & m = -1 \\ 0 & m \neq -1 \end{cases}$$

It follows that
$$\boxed{f(k) = \frac{1}{2\pi i}\oint_C z^{k-1}F(z)\, dz}$$

where the contour C encloses all the poles of $z^{k-1}F(z)$. This is called the Inversion Formula for Z-transform.

Now, since from the residue theorem

$$\oint_C z^{k-1}F(z)\, dz = 2\pi i \sum \text{Residues of } z^{k-1}F(z) \text{ at all its poles},$$

then

$$f(k) = \sum \text{Residues of } z^{k-1}F(z) \text{ at its poles}.$$

To find the Residues of $z^{k-1}F(z)$ at its poles, we can use one of the following expressions:

1. If $z^{k-1}F(z)$ has a simple pole at $z = a$, then

   Residue at the pole

   $$a = \lim_{z \to a} \{(z-a)z^{k-1}F(z)\}$$

2. If $z^{k-1}F(z)$ has a multiple pole of order n at $z = b$, then

   Residue at the pole

   $$b = \frac{1}{(n-1)!} \lim_{z \to b} \frac{d^{n-1}}{dz^{n-1}}\{(z-b)^n z^{k-1}F(z)\}.$$

Example: Use the Residue method to find the inverse Z-transform for

$$F(z) = \frac{z}{(z-1)^2} \ .$$

Solution: First, we multiply by z^{k-1} to get

$$z^{k-1}F(z) = \frac{z^k}{(z-1)^2}$$

The function has a double pole at $z = 1$, then

$$\text{Res} = \frac{1}{1!} \lim_{z \to 1} \frac{d}{dz}\left\{(z-1)^2 . \frac{z^k}{(z-1)^2}\right\}$$

$$= \lim_{z \to 1} \frac{d}{dz}\left\{z^k\right\} = \lim_{z \to 1} k \, z^{k-1} = k$$

Therefore $f(k) = k$. This can be verified directly from ***Table 1***.

Example: Use the Residue method to find the inverse Z-transform for

$$F(z) = \frac{z}{(z-1)(z-2)} \; .$$

Solution: First, we multiply by z^{k-1} to get

$$z^{k-1} F(z) = \frac{z^k}{(z-1)(z-2)}$$

The function has two simple poles at $z = 1$ and $z = 2$, then

$$(\text{Res at } z = 1) = \lim_{z \to 1} \left\{ (z-1) \cdot \frac{z^k}{(z-1)(z-2)} \right\} = -1 \, ,$$

$$(\text{Res at } z = 2) = \lim_{z \to 2} \left\{ (z-2) \cdot \frac{z^k}{(z-1)(z-2)} \right\} = 2^k \, ,$$

Therefore,

$$f(k) = \sum \text{Res} = 2^k - 1 \, .$$

Moreover, if this is the result of sampling at $T = 1$ second, then the corresponding continuous time signal is $f(t) = 2^t - u(t)$.

Example: What is the continuous time signal that when sampled at a sampling time $T = 1$ second will have a Z-transform given by $F(z) = \dfrac{1}{z^2(z-1)^2} \; .$

Solution: First, we multiply by z^{k-1} to get

$$z^{k-1} F(z) = \frac{z^{k-3}}{(z-1)^2}$$

The function has a double pole at $z = 1$, then

$$\text{Res} = \frac{1}{1!} \lim_{z \to 1} \frac{d}{dz} \left\{ (z-1)^2 \cdot \frac{z^{k-3}}{(z-1)^2} \right\}$$

$$= \lim_{z \to 1} \frac{d}{dz} \left\{ z^{k-3} \right\} = \lim_{z \to 1} (k-3) z^{k-1} = k-3$$

Therefore $f(k) = k - 3$.

Since this is the result of sampling at $T = 1$ second, then the corresponding continuous time signal is $f(t) = (t-3)u(t-3)$.

This can also be found using the shift theorem as follows:

We can write $F(z)$ in the form $F(z) = z^{-3} \dfrac{z}{(z-1)^2}$.

Now, from **Table 1**, we have $\mathbb{Z}\{ku(k)\} = \dfrac{z}{(z-1)^2}$,

then using the second shift theorem, we obtain

$$\mathbb{Z}\{(k-3)u(k-3)\} = z^{-3} \frac{z}{(z-1)^2}.$$

The result follows directly.

2.5. Exercises

1. Find the inverse Z-transform of each of the following functions. Put your answer in the simplest form:

(a) $F(z) = 3 + 2z^{-1} + 6z^{-4}$ Ans: $3\delta(k) + 2\delta(k-1) + 6\delta(k-4)$

(b) $F(z) = \dfrac{2z^2 - 3}{z^2 - 7z + 12}$ Ans: $7.25(4)^k - 5(3)^k - 0.25\delta(k)$

(c) $F(z) = \dfrac{10z(z+5)}{(z-1)(z-2)(z+3)}$ Ans: $(-3)^k + 14(2)^k - 15$

$(d)\,F(z)=\dfrac{10}{z^3-7z^2+12z}$ Ans: $0.625(4)^k-1.111(3)^k$
$+0.4861\delta(k)+0.833\delta(k-1)$

$(e)\,F(z)=\dfrac{4z^2-3z}{z^2+z+1/4}$ Ans: $\left[4(-0.5)^k-5k(-0.5)^k\right]u(k)$

$(f)\,F(z)=\dfrac{4z^2-3z+2}{z^2+z+1/4}$

Ans: $8\delta(k)+\left[-4(-0.5)^k+18k(-0.5)^k\right]u(k)$

$(g)\,F(z)=\dfrac{1+2z^{-1}}{1-2z^{-1}+z^{-2}}$ Ans: $(1+3k)\cdot u(k)$

$(h)\,F(z)=\dfrac{6}{\left(1+z^{-1}/4\right)\left(1+z^{-1}/2\right)}$

Ans: $\left[-6\left(\dfrac{-1}{4}\right)^k+12\left(\dfrac{-1}{2}\right)^k\right]u(k)$

$(i)\,F(z)=\dfrac{z^{-6}+z^{-7}}{1-z^{-1}}$ Ans: $u(k-6)+u(k-7)$

$(j)\,F(z)=2-\dfrac{3z}{z-4}$ Ans: $2\delta(k)-3(4)^k u(k)$

$(k)\,F(z)=\dfrac{1}{(z-1)^2}$ Ans: $(k-1)\cdot u(k-1)$

$(l)\,F(z)=\dfrac{-6z+1}{z^{50}\left(z^2+3z+2\right)}$ Ans: $-13(-2)^{k-51}+7(-1)^{k-51}$

$(m)\,F(z)=\dfrac{2z^2-1.5z}{z^2-1.5z+0.5}$ Ans: $\left(1+\left(\dfrac{1}{2}\right)^k\right)\cdot u(k)$

$(n)\,F(z)=\dfrac{1-z^{-1}/4}{1+z^{-1}/2}$ Ans: $\left(-\dfrac{1}{2}\right)^k\cdot u(k)-\dfrac{1}{4}\left(-\dfrac{1}{2}\right)^{k-1}\cdot u(k-1)$

$(o)\,F(z)=\dfrac{z^2-4z+14}{z^3-3z^2-10z+24}$

$$\text{Ans}:\left(-(2)^{k-1}+(-3)^{k-1}+4^{k-1}\right)\cdot u(k-1)$$

$(p)F(z)=\dfrac{4z^2-2z}{z^3-5z^2+8z-4}$

$\text{Ans}:2(1)^{k-1}\cdot u(k-1)+12(k-1)(2)^{k-2}\cdot u(k-2)+2(2)^{k-1}\cdot u(k-1)$

$(q)F(z)=\dfrac{2z^2-3z}{z^2+z+5/4}$

$$\text{Ans}:4.4722(1.118)^k\cos(1.1071k+1.1071)$$

$(r)F(z)=\dfrac{3z^3}{\left(z^2+1/4\right)(z-1/2)}$

$$\text{Ans}:1.5\left(\frac{1}{2}\right)^k+1.5\sqrt{2}\left(\frac{1}{2}\right)^k\cos\left(\frac{\pi}{2}\left(k-\frac{1}{2}\right)\right)$$

$(s)F(z)=\dfrac{5z(z+1)}{(z-1)(z^2+z+1)}$

$\text{Ans}:\dfrac{10}{3}(1)^{k-1}\cdot u(k-1)+\dfrac{5}{3}\cos\dfrac{2\pi}{3}(k-1)\cdot u(k-1)+\dfrac{5}{\sqrt{3}}\sin\dfrac{2\pi}{3}(k-1)\cdot u(k-1)$

$(t)F(z)=\dfrac{z^2+4z-5}{\left(z-\dfrac{1}{2}+j\dfrac{1}{4}\right)\left(z-\dfrac{1}{2}-j\dfrac{1}{4}\right)\left(z-\dfrac{1}{2}\right)}$

$$\text{Ans}:32\delta(k)-88\left(\frac{1}{2}\right)^k+88.1(0.559)^k\cos\left(0.464k-0.882\right)$$

2. Use long division to show that:

$(a)Z^{-1}\left\{\dfrac{2z^2}{z^2-1/4}\right\}=\left(\dfrac{1}{2}\right)^k+\left(-\dfrac{1}{2}\right)^k \qquad k=0,1,2,3,......$

$(b)Z^{-1}\left\{\dfrac{8z}{(z-1)^2}\right\}=8k \qquad k=0,1,2,3,...$

59

$(c)Z^{-1}\left\{\dfrac{z\,(z-1/2)}{z^2-z+1}\right\}=\cos\left(\dfrac{1}{2}\right)k \qquad k=0,1,2,3,\ldots$

3. The Z-transform of a signal $f(k)$ is given by

$$F(z)=\dfrac{2z^6-z^5+3z^3+2z^2}{z^7+2z^6+z^5+z^4+0.5}$$

Compute $f(3)$. $\qquad\qquad\qquad\qquad\qquad$ $\left[\text{Ans}:f(3)=8\right]$

Hint: Since we are not interested in getting a closed form of $f(k)$, it is better to use the long division method and stop at the coefficient of z^{-3}.

Chapter Three

Applications of the $\mathbb{Z}$ -Transform

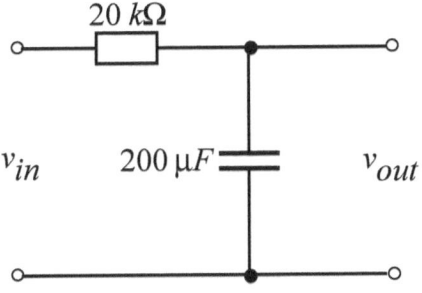

Refaat El Attar

Chapter 3

Applications of the $\mathbb{Z}$ -Transform

3.1. Introduction

Many applications of transforms in general are found in engineering and computer science. In particular Z-transform is a mathematical tool that helps solve several practical problems in difference equations and digital filtering among others. Here, we study some applications of the Z-transform.

3.2. Difference Equations

Difference equations are the discrete equivalent of differential equations. Methods for solving difference equations are very similar to those for solving differential equations. As we know, Laplace Transform is used to solve differential equations where the variables involved are of a continuous nature in general. The Z-transform plays the same role when the variables are discrete in nature or as the results of a sampling process.

Consider the simple difference equation $y(k+1) - y(k) = 4$. The dependent variable is y while the independent variable is k. A solution of the difference equation is obtained by finding the dependent variable y for each value of the independent variable k. Many sequences may satisfy a difference equation, then the general solution is a linear combination of all these sequences that satisfy the difference equation.

The order of a difference equation is the difference between the highest and lowest argument of y. For example

$$y(k+2) - y(k+1) - 2y(k) = 2k$$

is second order, while

$$y(k+1) - 4y(k-1) + y(k-2) = 3^{k}$$

is third order.

The dependent variable is some equations occurs only once, then they are of zero order and are called *non-recursive* differential equations. This is because the calculation of the value of the dependent variable does not require the knowledge of its previous values. Difference equations of order greater then zero are called *recursive* difference equations.

The method of solution can be summarized in the following steps:

1. Transform all the parts of the given difference equation to the Z-transform domain.

2. Substitute with the initial conditions.

3. Solve the resulting algebraic equation to get $F(z)$.

4. Find the inverse Z-transform for $F(z)$ and now you get the solution of the difference equation.

This is illustrated in the following examples.

Example: Using Z-transform, find the solution of the difference equation $y(k+1) - 4y(k) = \delta(k)$, $y(0) = 0$.

Solution: Taking the Z-transform of both sides, we get

$$zY(z) - zy(0) - 4Y(z) = 1$$

Substituting with initial condition, we get

$$(z - 4)Y(z) = 1, \text{ or}$$

$$Y(z) = \frac{1}{z-4} = z^{-1}\frac{z}{z-4},$$

then from ***Table 1***, we obtain

$$y(k) = 4^{(k-1)}u(k-1).$$

Example: Using Z-transform, find the solution of the difference equation

$$y(k+2) - y(k+1) + \frac{2}{9}y(k) = u(k), \quad y(0) = 1, y(1) = 1.$$

Solution: Taking the Z-transform of both sides, we get

$$z^2\left[Y(z)-y(0)-y(1)z^{-1}\right]-z\left[Y(z)-y(0)\right]+\frac{2}{9}Y(z)=\frac{z}{z-1}$$

Substituting with initial conditions:

$$\left(z^2-z+\frac{2}{9}\right)Y(z)=\frac{z}{z-1}+z^2=z\cdot\frac{z^2-z+1}{z-1}$$

Rearranging we get:

$$Y(z)=z\cdot\frac{z^2-z+1}{(z-1)\left(z-\frac{1}{3}\right)\left(z-\frac{2}{3}\right)}$$

Performing partial fraction decomposition, we obtain

$$Y(z)=z\cdot\left[\frac{9/2}{z-1}+\frac{7/2}{z-1/3}-\frac{7}{z-2/3}\right]$$

$$Y(z)=\frac{9}{2}\cdot\frac{z}{z-1}+\frac{7}{2}\cdot\frac{z}{z-1/3}-\frac{7z}{z-2/3}$$

then from **Table 1**, we obtain

$$y(k)=\frac{9}{2}u(k)+\frac{7}{2}\left(\frac{1}{3}\right)^k u(k)-7\left(\frac{2}{3}\right)^k u(k).$$

Example: Using Z-transform, find the solution of the difference equation

$$y(k)-\frac{3}{4}y(k-1)+\frac{1}{8}y(k-2)=2\sin(\frac{\pi k}{2})$$,

$$y(-1)=2,\ y(-2)=4.$$

Solution: Taking the Z-transform of both sides, we get

$$Y(z)-\frac{3}{4}z^{-1}\left[Y(z)+2z\right]+\frac{1}{8}z^{-2}\left[Y(z)+4z^2+2z\right]=\frac{2z}{z^2+1}$$

65

$$\left(1-\frac{3}{4}z^{-1}+\frac{1}{8}z^{-2}\right)Y(z)=1-\frac{1}{4}z^{-1}+\frac{2z}{z^2+1}$$

$$Y(z)=\frac{z^2-\frac{1}{4}z}{z^2-\frac{3}{4}z+\frac{1}{8}}+\frac{2z^3}{\left(z^2+1\right)\cdot\left(z^2-\frac{3}{4}z+\frac{1}{8}\right)}$$

Performing Partial fraction decomposition, we obtain

$$Y(z)=\frac{z}{z-\frac{1}{2}}+\frac{8}{5}\cdot\frac{z}{z-\frac{1}{2}}-\frac{8}{17}\cdot\frac{z}{z-\frac{1}{4}}+\frac{\frac{112}{85}z^2-\frac{96}{85}z}{z^2+1}$$

This problem was solved before in the previous section

$$y(k)=\frac{13}{5}\left(\frac{1}{2}\right)^k-\frac{8}{17}\left(\frac{1}{4}\right)^k+\frac{112}{85}\sin\frac{\pi}{2}k-\frac{96}{85}\cos\frac{\pi}{2}k; \quad k\geq 0$$

Example: Using Z-transform, find the solution of the difference equation:

$$f(k+2)-5f(k+1)+6f(k)=4^k, \quad f(0)=0, f(1)=1.$$

Solution: Taking the Z-transform of both sides, we get

$$z^2F(z)-\{z^2f(0)+zf(1)\}-5\{zF(z)-zf(0)\}+6F(z)=\frac{z}{z-4}$$

Substituting for $f(0)=0$, $f(1)=1$, and solving for $F(z)$, we obtain

$$F(z)=\frac{z}{(z-2)(z-4)}.$$

Using partial fraction decomposition, we obtain

$$\frac{F(z)}{z}=-\frac{1}{2}\left\{\frac{1}{z-2}-\frac{1}{z-4}\right\}, \text{ or}$$

$$F(z) = -\frac{1}{2}\left\{ \frac{z}{z-2} - \frac{z}{z-4} \right\},$$

then from **Table 1**, we have $f(k) = \frac{1}{2}(4)^k - \frac{1}{2}(2)^k$, the sequence of which is $f(k) = \{0, 1, 6, ...\}$.

Example: Using Z-transform, find the solution of the difference equation:

$$2y_{k+2} - 7y_{k+1} + 3y_k = 8 , \quad y_0 = -1, \ y_1 = 0.$$

Solution: Taking the Z-transform of both sides, we get

$$2z^2 Y(z) - 2\{z^2 y_0 + zy_1\} - 5\{zY(z) - zy_0\} + 3Y(z) = \frac{8z}{z-1}$$

Substituting for $y_0 = -1$, $y_1 = 0$, we get

$$(2z^2 - 7z + 3)Y(z) = \frac{8z}{z-1} - 2z^2 + 7z$$

Solving for $Y(z)$, we obtain $Y(z) = \dfrac{-z(2z^2 - 9z - 1)}{(z-1)(2z-1)(z-3)}$.

Using partial fraction decomposition, we obtain

$$\frac{Y(z)}{z} = \frac{-4}{z-1} + \frac{4}{2z-1} + \frac{1}{z-3}, \text{ or}$$

$$Y(z) = \frac{-4z}{z-1} + \frac{2z}{z-\dfrac{1}{2}} + \frac{z}{z-3},$$

then from **Table 1**, we obtain

$$y(k) = -4 + 2\left(\frac{1}{2}\right)^k + (3)^k,$$

the sequence of which is $f(k) = \{-1, 0, 5.5, ...\}$.

Example: Using Z-transform, find the solution of the difference equation

$$y_{k+2} - 2y_{k+1} + 4y_k = 0, \quad y_0 = -0 \ y_1 = 1.$$

Solution: Taking the Z-transform of both sides, we get

$$z^2 Y(z) - \{z^2 y_0 + zy_1\} - 2\{zY(z) - zy_0\} + 4Y(z) = 0$$

Substituting for $y_0 = 0$, $y_1 = 1$, we get

$$(z^2 - 2z + 4)Y(z) = z$$

Solving for $Y(z)$, we obtain

$$Y(z) = \frac{z}{z^2 - 2z + 4}.$$

To get the inverse transform, we use the long division method to obtain the sequence

$$f(k) = \{0, 1, 2, 0, -8, \ldots\}.$$

Example: The Fibonacci sequence is obtained by starting with $a(0) = a(1) = 1$ and then setting $a(2) = a(1) + a(0) = 2$, $a(3) = a(2) + a(1) = 3$, $a(4) = \quad a(3) + a(2) = 5$, etc.; the sequence is continued with the use of the general formula $a(k+2) = a(k+1) + a(k)$. Find the general term in the Fibonacci sequence.

Solution: At first, the difference equation that describes our problem:

$$a(k+2) = a(k+1) + a(k)$$

And the initial conditions are: $a(0) = 1$ and $a(1) = 1$

Taking the Z transform of the difference equation:

$$\left(z^2 A(z) - z^2 a(0) - za(1)\right) - \left(zA(z) - za(0)\right) - A(z) = 0$$

$$(z^2 - z - 1)A(z) = z^2 a(0) - za(0) + za(1) = z^2$$

$$A(z) = \frac{z^2}{z^2 - z - 1} = z \cdot \frac{z}{\left(z - \frac{1+\sqrt{5}}{2}\right)\left(z - \frac{1-\sqrt{5}}{2}\right)}$$

Performing partial fraction expansion:

$$A(z) = \frac{5+\sqrt{5}}{10}\left(\frac{z}{z - \frac{1+\sqrt{5}}{2}}\right) + \frac{5-\sqrt{5}}{10}\left(\frac{z}{z - \frac{1-\sqrt{5}}{2}}\right)$$

Taking the inverse Z transform:

$$a(k) = \frac{5+\sqrt{5}}{10}\left(\frac{1+\sqrt{5}}{2}\right)^k + \frac{5-\sqrt{5}}{10}\left(\frac{1-\sqrt{5}}{2}\right)^k$$

This represents the law for the general term in the Fibonacci sequence.

Example: A symmetrical ladder network is shown in Figure 8, write the difference equation describing the network behavior. Specify the boundary conditions and solve for the current in any loop. Assume that each resistor is $1\,\Omega$.

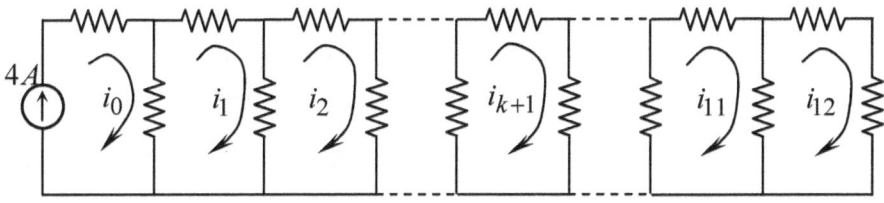

Figure 8.

Solution: Applying Kirchoff's law for the $(k+1)th$ loop, we get the difference equation

$$3Ri_{k+1} - Ri_k - Ri_{k+2} = 0$$

and since $R = 1\,\Omega$, we get

$$3i_{k+1} - i_k - i_{k+2} = 0$$

The boundary conditions are

$$i_0 = 4, \ 3i_{12} - i_{11} = 0.$$

Taking the Z-transform, we obtain

$$\{3zI(z) - 3zi_0\} - I(z) - \{z^2 I(z) - z^2 i_0 - zi_1\} = 0.$$

Substituting for i_0 and solving for $I(z)$, we obtain

$$I(z) = \frac{4z^2 - 12z + i_1 z}{z^2 - 3z + 1} = \frac{4z\left(z - \dfrac{3}{2}\right)}{z^2 - 2z\left(\dfrac{3}{2}\right) + 1} - \frac{(6 - i_1)z}{z^2 - 2z\left(\dfrac{3}{2}\right) + 1}$$

Taking the inverse transform using the entries in *Table* **1**, we get

$$i_k = 4\cosh \alpha k - \frac{6 - i_1}{\sinh \alpha}\sinh \alpha k ,$$

Where

$$\cosh \alpha = \frac{3}{2}, \text{ or } \alpha = 0.9624 .$$

To determine i_1, we use the second boundary condition

$$3\left[4\cosh 12\alpha - \frac{6 - i_1}{\sinh \alpha}\sinh 12\alpha \right]$$
$$= \left[4\cosh 11\alpha - \frac{6 - i_1}{\sinh \alpha}\sinh 11\alpha \right].$$

Solving for i_1, we get $i_1 = 1.52$. Finally, the current in the $(k+1)th$ loop is

$$i_k = 4\cosh(0.9624k) - \frac{4.48\sinh(0.9624k)}{\sinh(0.9624)} .$$

3.3. The Transfer Function

This function describes the relationship between the digital output $Y(z)$ and the digital input $X(z)$ of a block in a system when all initial conditions are zeros. It is given by

$$G(z) = \frac{Y(z)}{X(z)}$$

This transfer function can be used to describe the operation of blocks in such systems as digital control systems and digital filters.

Example: Determine the discrete transfer function for the electric circuit shown in Figure 9, when its input is sampled at the rate of $10\ Hz$.

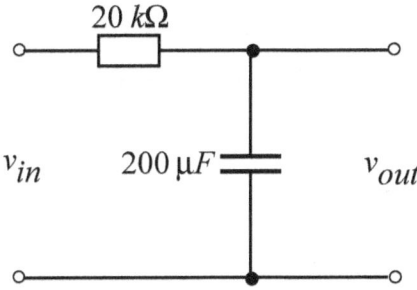

Figure 9. Example

Solution: Applying Kirchoff's law, we get the equation

$$Ri(t) + \frac{1}{C}\int i(t)\ dt = v_{in}(t)$$

$$v_{out}(t) = \frac{1}{C}\int i(t)\ dt$$

Taking Laplace transform, we get

$$RI(s) + \frac{1}{Cs}I(s) = V_{in}(s) \text{ and } V_{out}(s) = \frac{1}{Cs}I(s).$$

Therefore $RCsV_{out}(s) + V_{out}(s) = V_{in}(s)$.

The transfer function is then

$$G(s) = \frac{V_{out}(s)}{V_{in}(s)} = \frac{1}{1 + RCs} = \frac{1/4}{s + 1/4}$$

Then from **Table 1**, we get

$$G(s) = \frac{V_{out}(s)}{V_{in}(s)} = \frac{1}{1 + RCs} = \frac{1/4}{s + 1/4}$$

and since $R = 1\,\Omega$, we get $3i_{k+1} - i_k - i_{k+2} = 0$

The boundary conditions are $i_0 = 4$, $3i_{12} - i_{11} = 0$.

Taking the Z-transform, we obtain

$$\{3zI(z) - 3zi_0\} - I(z) - \{z^2 I(z) - z^2 i_0 - zi_1\} = 0.$$

Substituting for i_0 and solving for $I(z)$, we obtain

$$I(z) = \frac{4z^2 - 12z + i_1 z}{z^2 - 3z + 1}$$

$$= \frac{4z(z - 3/2)}{z^2 - 2z(3/2) + 1} - \frac{(6 - i_1)z}{z^2 - 2z(3/2) + 1}$$

Taking the inverse transform using the entries in **Table 1**, we get

$$i_k = 4\cosh \alpha k - \frac{6 - i_1}{\sinh \alpha} \sinh \alpha k,$$

where $\cosh \alpha = \dfrac{3}{2}$, or $\alpha = 0.9624$.

To determine i_1, we use the second boundary condition

$$3\left[4\cosh 12\alpha - \frac{6-i_1}{\sinh\alpha}\sinh 12\alpha\right]$$
$$= \left[4\cosh 11\alpha - \frac{6-i_1}{\sinh\alpha}\sinh 11\alpha\right].$$

Solving for i_1, we get $i_1 = 1.52$

Finally, the current in the $(k+1)th$ loop is

$$i_k = 4\cosh(0.9624k) - \frac{4.48\sinh(0.9624k)}{\sinh(0.9624)}.$$

Example: The difference equation of a digital filter is given in the following equation. We want to find the mathematical description (transfer function) of this filter $H(z)$.

$$y_k = 1.85y_{k-1} - 0.868y_{k-2} + 0.00477x_{k-1} + 0.00455x_{k-2}$$

Solution: Taking the Z-transform of both sides in the equation, we get the following equation.

$$Y(z) = Y(z).[1.85z^{-1} - 0.868z^{-2}]$$
$$+ X(z).[0.00477z^{-1} + 0.00455z^{-2}]$$

Solving for $H(z)$ gives the following equation

$$H(z) = \frac{Y(z)}{X(z)} = \frac{0.00477z^{-1} + 0.00455z^{-2}}{1 - 1.85z^{-1} + 0.868z^{-2}}$$

$$H(z) = \frac{0.00477z + 0.00455}{z^2 - 1.85z + 0.868}.$$

Example: Given the following transfer function of a digital system, determine the difference equation that describes the system in the time domain.

$$H(z) = \frac{0.25z^2 - 0.5z + 0.25}{z^2 - 0.95z + 0.75}$$

73

Solution: First let's divide numerator and denominator by z^2, as shown in the following equation

$$H(z) = \frac{Y(z)}{X(z)} = \frac{0.25 - 0.5z^{-1} + 0.25z^{-2}}{1 - 0.95z^{-1} + 0.75z^{-2}}$$

Then cross-multiply the preceding equation to get the following equation.

$$Y(z)[1 - 0.95z^{-1} + 0.75z^{-2}] = X(z)[0.25 - 0.5z^{-1} + 0.25z^{-2}]$$

Now multiplying through each term in the brackets gives the following equation.

$$Y(z) = 0.95\ Y(z)z^{-1} - 0.75\ Y(z)z^{-2}$$
$$+ 0.25X(z) - 0.5X(z)z^{-1} + 0.25X(z)z^{-2}$$

Finally the inverse Z-transform of the signals can be taken by applying the shifting property

$$y_k = 0.95\ y_{k-1} - 0.75\ y_{k-2} + 0.25x_k - 0.5x_{k-1} + 0.25x_{k-2}$$

3.4. Exercises

1. Solve the following difference equations:

$(a)\ y(k) - ay(k-1) = 2\delta(k),\ y(0) = 0$ \qquad Ans: $y(k) = 2a^k u(k)$

$(b)\ y(k+2) - 5y(k+1) + 6y(k) = 0,\quad y(0) = 0, y(1) = 1$

$$\text{Ans}: y(k) = \left(3^k - 2^k\right) \cdot u(k)$$

$(c)\ 3y(k+2) - 5y(k+1) + 2y(k) = 0,\quad y(0) = 1, y(1) = 0$

$$\text{Ans}: y(k) = 3(2/3)^k - 2$$

$(d\,)y\,(k\,+2)-4y\,(k\,+1)+4y\,(k\,)=0, \quad y\,(0)=1, y\,(1)=4$

$$\text{Ans} : y\,(k\,)=2^k\,(k\,+1)$$

$(e\,)y\,(k\,+2)-5y\,(k\,+1)+6y\,(k\,)=2k\,+1, \quad y\,(0)=0, y\,(1)=1$

$$\text{Ans} : y\,(k\,)=\frac{5}{2}(3)^k\,-5(2)^k\,+k\,+\frac{5}{2}$$

$(f\,)y\,(k\,+2)-6y\,(k\,+1)+5y\,(k\,)=2^k\,, \quad y\,(0)=0, y\,(1)=0$

$$\text{Ans} : y\,(k\,)=\frac{1}{4}+\frac{1}{12}(5)^k\,-\frac{1}{3}(2)^k$$

$(g\,)y\,(k\,+2)-2y\,(k\,+1)+y\,(k\,)=\sin\frac{\pi}{2}k\,, \quad y\,(0)=0, y\,(1)=1$

$$\text{Ans} : y\,(k\,)=\frac{3}{2}k\,+\frac{1}{2}\cos\frac{\pi}{2}k\,-\frac{1}{2}$$

$(h)y\,(k\,+2)=\frac{1}{3}y\,(k\,+1)+\frac{1}{3}y\,(k\,)+x\,(k\,+1)-x\,(k\,),$

$y\,(-1)=0, y\,(-2)=-1, \qquad x\,(k\,)=\left[3+4\left(\frac{1}{4}\right)^k\right]u\,(k\,)$

$$\text{Ans} : y\,(k\,)=0.7889(0.7676)^k\,-9.5927(-0.4343)^k\,+8.4706(0.25)^k$$

$(i\,)y\,(k\,+3)-2y\,(k\,+2)-2y\,(k\,+1)=2x\,(k\,+2)-x\,(k\,),$

$y\,(-1)=1, y\,(-2)=-2, y\,(-3)=0, \qquad x\,(k\,)=3\delta(k\,-1)$

$$\text{Ans} : y\,(k\,)=5.6629\left(\sqrt{2}\right)^k\,\cos\left(\frac{\pi}{4}k\,-2.4186\right)+2.25\delta(k\,)$$

$$-1.5\delta(k\,-1)-1.5\delta(k\,-2)$$

$(j)y(k+2) = -y(k+1) - \frac{1}{2}y(k) + 4x(k+2) - 3x(k+1),$

$y(-1) = 2, y(-2) = 0, \qquad x(k) = (1/2)^k u(k)$

$$\text{Ans}: y(k) = 8.2462\left(\frac{\sqrt{2}}{2}\right)^k \cos(0.7854k - 0.245) - 2(0.5)^k$$

$(k)ky(k+1) + y(k+1) = y(k), \quad y(0) = 1 \qquad \text{Ans}: y(k) = \frac{1}{k!}$

2. Given the following difference equation of a discrete-time system, find the transfer function $H(z)$:

$(a)y(k+2) + 2y(k+1) - 4y(k) = 8x(k+1) - 6x(k)$
$$\text{Ans}: H(z) = \frac{8z - 6}{z^2 + 2z - 4}$$

$(b)y(k+2) - y(k+1) + y(k) = x(k+3) - x(k+2) + 4x(k+1) + x(k)$
$$\text{Ans}: H(z) = \frac{z^3 - z^2 + 4z + 1}{z^2 - z + 1}$$

$(c)y(k+3) = x(k+3) - 2x(k+2) + 4x(k+1) - 2x(k)$
$$\text{Ans}: H(z) = \frac{z^3 - 2z^2 + 4z - 2}{z^3}$$

$(d)y(k) = -2y(k-1) + 3x(k) + x(k-1) \qquad \text{Ans}: H(z) = \frac{3 + z^{-1}}{1 + 2z^{-1}}$

$(e)2y(k) + y(k-1) = -3y(k-2) + 4x(k) - 2x(k-2)$
$$\text{Ans}: H(z) = \frac{4 - 2z^{-2}}{2 + z^{-1} + 3z^{-2}}$$

3. Given the following transfer function $H(z)$ of a discrete-time system, write the difference equation that describes the system:

$(a)H(z) = \frac{3 + 2z^{-1} + z^{-2}}{1 - 4z^{-1} + 5z^{-2}}$ $\begin{bmatrix} \text{Ans}: y(k) = 4y(k-1) - 5y(k-2) \\ + 3x(k) + 2x(k-1) + x(k-2) \end{bmatrix}$

$b)H(z) = \dfrac{a_0 + a_1 z^{-1}}{1 - b_1 z^{-1}}$ $\text{Ans}: y(k) = b_1 y(k-1) + a_0 x(k) + a_1 x(k-1)$

$(c)H(z) = \dfrac{3z + 5}{2z^2 - 5z + 4}$ $\left[\begin{array}{l}\text{Ans}: y(k) = 2.5y(k-1) - 2y(k-2) \\ \qquad\qquad + 1.5x(k-1) + 2.5x(k-2)\end{array}\right]$

$(d)H(z) = \dfrac{5z^2 + 4z}{z^2 + \dfrac{3}{2}z + \dfrac{1}{2}}$

$$\text{Ans}: y(k) = -\frac{3}{2}y(k-1) - \frac{1}{2}y(k-2) + 5x(k) + 4x(k-1)$$

www.ingramcontent.com/pod-product-compliance
Lightning Source LLC
Chambersburg PA
CBHW022128170526
45157CB00004B/1797